DISEASES OF SPICE CROPS

NIPA GENX ELECTRONIC RESOURCES & SOLUTIONS P. LTD.
New Delhi-110 034

About the Authors

Dr. L. Darwin Chrisdhas Henry obtained his B.Sc. Degree in Botany from Madras Christian College, Chennai. He continued his under graduate and post graduate degree courses in Agriculture at Allahabad Agricultural Institute. He joined as an Assistant Professor in Plant Pathology, Faculty of Agriculture, Annamalai University. He has completed his doctorate specialising in the field of edible mushrooms. He has published a number of research articles in National and International Journals. He has authored a book entitled 'Crop Diseases' published by NIPA, New Delhi and also a book entitled 'Plant Ecology' published by M/S. Agri-Biovet Press, New Delhi. He is also a coauthor of a book entitled 'Illustrated Plant Pathology' published by NIPA, New Delhi. Apart from his wide exposure to different agricultural crops, he has specialised in the cultivation of different species of edible mushrooms.

H. Lewin Devasahayam after completion of his post graduate education from the Banaras Hindu University, Varanasi joined the Tamil Nadu Agricultural University in 1960 and served the university for 30 years in various capacities and retired as Associate Professor of Plant Pathology in 1990. He has published more than 50 research articles in various leading Indian and foreign jounals. He is the author of the books entitled 'Practical manual of Entomology', 'Elemets of Entomoly' 'Illutrated Plant Pathology' and co-author of the book 'Crop diseases', all the above four books have been published by NIPA, New Delhi. He is also the co-author of the book entitled 'Plant Ecology' published by M/S. Agri-Biovet Press, New Delhi.

DISEASES OF SPICE CROPS

L. Darwin Christdhas Henry
Associate Professor
Department of Plant Pathology
Annamalai University, Chidambaram
Tamil Nadu

H. Lewin Devasahayam
(Retd.) Associate Professor
Department of Plant Pathology
Tamil Nadu Agricultural University
Coimbatore, Tamil Nadu

NIPA GENX ELECTRONIC RESOURCES & SOLUTIONS P. LTD.
New Delhi-110 034

NIPA GENX ELECTRONIC RESOURCES & SOLUTIONS P. LTD.

101,103, Vikas Surya Plaza, CU Block
L.S.C.Market, Pitam Pura, New Delhi-110 034
Ph : +91 11 27341616, 27341717, 27341718
E-mail:newindiapublishingagency@gmail.com
www: www.nipabooks.com

For customer assistance, please contact
Phone: + 91-11-27 34 17 17 Fax: + 91-11- 27 34 16 16
E-Mail: feedbacks@nipabooks.com

ISBN: 978-93-91383-97-8

Composed and Designed by NIPA.

Dr. M. Ganapathy
Dean
Faculty of Agriculture
Annamalai University

Foreword

India is the largest producer and exporter of spices, spice oils and oleoresins which play an important role in our country's economy by earning valuable foreign exchange, providing direct and indirect employment to many people and also by supporting a number of horticultural based industries.

With limitations in many input parameters, such as cultivable land, water, fertilizer etc., it is imperative that measures have to be taken to increase crop production with the available resources. Among the various constraints faced in crop husbandry, plant diseases play a very important role. Under favorable conditions some diseases may cause even cent per cent loss in commercial crops. This ultimately results in huge loss of marketable produce and monetary loss. To adopt various strategies for the control of pathogens, one should have some basic knowledge about the symptoms produced by the pathogens, their life cycle, mode of survival and spread, and the stage at which the host is most vulnerable to attack by the pathogens etc. To save the crops from these microscopic pathogens, knowledge of diseases and measures to control them are of vital importance.

In this book, the authors have given detailed account of the major diseases of important spice crops, along with their integrated management practices. Further, the mycological aspects of important genera of fungi causing the diseases along with the life cycle of the fungal pathogens have also been included. The text is substantiated with photos revealing the symptoms of diseases, and many fine, hand-drawn illustrations, which is a highlight of the book. This book will be of immense use to the students of Agriculture, Horticulture and Forestry in the Agricultural Universities and Institutions and also as a reference book to those working in the State Agricultural, Horticultural and Forestry Departments.

I congratulate the authors Dr. L. Darwin Christdhas Henry and Thiru. H. Lewin Devasahayam for their great effort in bringing out this very useful book.

Annamalai Nagar
26.11.2021

Dr. M. Ganapathy
Dean
Faculty of Agriculture
Annamalai University

Preface

The history of spices is almost as old as human civilization. It is a history of lands discovered, empires built and brought down, wars won and lost, treaties signed and flouted, flavors sought and offered and the rise and fall different religious practices and beliefs. Spices were among the most valuable items of trade in the ancient and medieval times. Around 3500 BC the ancient Egyptians were using spices for flavoring food, in cosmetics and for embalming their dead. The uses of spices spread rapidly through the Middle East to Eastern Mediterranean and Europe. Spices have been closely connected to magic, culture, tradition, preservation, medicine and embalming since early in human history. Spices played a vital role in India's external trade with Mesopotamia, China, Egypt and Arabia along with perfumes and textiles as far back as 7000 years ago, way before the Greek and Roman civilization. Subsequently different countries worldwide have monopolized spice production and trade at different times. Now spice cultivation has become quite common and almost all countries of the world are producing spices depending upon the climatic conditions of the countries. The climate of India is quite suitable for the cultivation of most of the spices and so India has become a forerunner of spice production and trade. India is earning millions of dollars in foreign exchange as a result of spice trade.

A spice is a seed, fruit, root, bark or any other plant parts. Spices give aroma, color, flavor, taste and texture to food. Each spice has its unique, specific compounds that impart these sensual qualities. Spices are often used in medicines, religious rituals, cosmetics and perfume production.

A spice may be available in several forms, fruits, whole, dried or ground. Generally spices are dried whole and stored. They may be ground into a powder form for convenience. A whole dried spice has the longest shelf life and so it can be stored in bulk quantities. A fresh spice, such as ginger is usually more flavorful than in its dried form. But fresh spices are more expensive and have a much shorter shelf life. Some spices, such as turmeric are not always available fresh. Small spice seeds, such as fennel and mustard seeds are often used whole and in powder form.

The flavor of a spice is derived from volatile oils that get oxidized or evaporated when exposed to air. Grinding a spice greatly increases its surface area and increases its rate of oxidation and evaporation. Thus the flavor is retained by storing a spice whole and grinding it when needed, as the flavor life of a ground spice is much shorter.

Some flavor elements in spices are soluble in water, while many are soluble in oil or fat. The flavor from a spice takes a long time to infuse into the food and so spices are added early in food preparations. On the other hand, herbs are added late in the food preparation.

Because spices impart strong flavors, they are used in small quantities. Most spices and herbs possess antioxidant property owing primarily to the presence of phenol compounds, especially 'flavanoids' that enhances absorption of other nutrients. Cumin and ginger have been found to induce high antioxidant activity.

Many of the spices were known for their therapeutic properties well before culinary uses. Modern medical science has shown that many of the spices do possess remarkable health benefits. A few are known to cure some illness, while some are found to ameliorate the severity of certain diseases. Although spices have therapeutic properties and can cure many diseases and minor ailments, they are susceptible to many fungal, bacterial and viral diseases. The pathogens causing the diseases may attack the foliage, stems, roots, fruits, seeds, rhizomes, bulbs and any other part of the plants and cause severe damage to the crops resulting in huge loss of marketable produce. Some pathogens may even kill the affected plants.

In this book the diseases affecting 29 spice crops have been dealt with. A short description about each crop and the produce as well as the symptoms and possible control measures of the diseases attacking the crop are described in detail. A number of color photographs of the crops and produce and symptoms manifested by the diseases have been given. Further general characters of different genera of fungi causing the diseases along with 32 figures depicting the life cycle of the fungal pathogens have also been included.

All the photographs in the book have been taken from the Internet and the authors are highly thankful and express their sincere gratitude to those who have contributed the photographs in public interest.

The book will be of immense use to both undergraduate and postgraduate students pursuing Agriculture Degree Courses, as well as to the elite Public interested in the cultivation of spice crops.

The authors are extremely thankful to the Authorities, Annamalai University, Chidambaram, Tamil Nadu for having accorded permission to publish the book.

The authors are also highly thankful to Dr. M. Ganapathy Dean, Faculty of Agriculture, Annamalai University for being kind enough to give the foreword to the book.

The authors wish to express their appreciation and sincere thanks to M/S NIPA GENX Electronic Resources & Solutions P. Ltd. New Delhi-110 034 for the excellent manner in which the book has been brought out.

L. Darwin Christdhas Henry
H. Lewin Devasahayam

Contents

List of Figures

List of Fungicides Recommended for Spraying on Spice Crops

Chemical name of fungicide	Trade names	Dosage (% of formulation)	Dosage (g/litre of water)
Bordeaux mixture	-	1.00%	-
Captafol	Difolatan, Captafol, Foltaf, Sanspor, Difosan etc.	0.15%	1.5 g / litre of water
Captan	Captan 50, Esso fungicide 406, Orthocide, Merpan, Hexacap, Vancide etc.	0.15%	1.5 g / litre of water
Carbendazim	Bavistin, Derosal, Bengard, Zoom, Tagstin, MBC, Agrozim etc.	0.1%	1.0 g / litre of water
Chlorothalonil	Daconil, Bravo, Termil, Kavach, Speldrum etc.	0.15%	1.5 g / litre of water
Copper oxychloride	Fytolon, Cupramar, Blitox, Cuprocide, Blue copper, Parrycop etc.	0.25%	2.5 g / litre of water
Dinocap	Karathane, Crotothane, Arathane, Capryl, Mildex etc.	0.15%	1.5 g / litre of water
Fosetyl-Al	Aliette etc.	0.10%	1.0 g / litre of water
Hexaconazole	Contaf etc	0.15%	1.5 g / litre of water
Iprodione	Rovral, Chipco 26019, Glycophene etc.	0.15%	1.5 g / litre of water
Mancozeb	Mancozeb, Dithane M 45, Manzate 200, Pencozeb etc Ridomil, Acylon,	0.20%	2.0 g / litre of water

Metalaxyl	Apron etc.	0.10%	1.0 g / litre of water
Propiconazole	Tilt etc.	0.10%	1.0 g / litre of water
Thiophanate methyl	Topsin M, Cercobin M, Enovit methyl etc.	0.10%	1.0 g / litre of water
Thiram	Thiram, Thiride, Arasan, Hexathir, Panoram, Tersan etc.	0.20%	2.0 g / litre of water
Triazoles	Bayleton, Bayton, Propiconazole, Hexaconazole	0.10%	1.0 g / litre of water
Wettable sulphur	Thiovit, Cosan, Sulfex, Elosan etc.	0.40%	4.0 g / litre of water

Classification of Fungal Pathogens of Spice Crops (Division; Class; Order; Family)

Pathogen	Classification
Albugo candida	Oomycota; Oomycetes; Peronosporales; Albuginaceae
Alternaria alternata *Alternaria brassicae* *Alternaria burnsii* *Alternaria* (*Macrosporium*) *porri*	Ascomycota; Dothideomycetes; Pleosporales; Pleosporaceae
Aspergillus niger	Ascomycota; Eurotiomycetes; Eurotiales; Trichocomataceae
Botrytis cinerea	Ascomycota; Leotiomycetes; Helotiales; Sclerotiniaceae
Bremia lactucae	Oomycota; Oomycetes; Peronosporales; Peronosporaceae
Ceratocystis fimbriata	Ascomycota; Sordariomycetes; Microascales; Ceratocystidaceae
Cercospora ocimicola *Cercospora traversiana* *Cercospora zingiberi*	Ascomycota; Dothideomycetes; Capnodiales; Mycosphaerellaceae
Colletotrichum zingiberis *Colletotrichum gloeosporioides* *Colletotrichum necator* *Colletotrichum capsici* *Colletotrichum vanillae*	Ascomycota; Sordariomycetes; Glomerellales; Glomerellaceae
Corticium salmonicolor/ *Erythricium salmonicolor*	Basidiomycota; Agaricomycetes; Corticiales; Corticiaceae
Cryphonectria cubensis	Ascomycota; Sordariomycetes; Diaporthales; Cryphonectriaceae
Curvularia andropogonis	Ascomycota; Dothideomycetes; Pleosporales; Pleosporaceae
Cylindrocladium quinquiseptatum	Ascomycota; Sordariomycetes; Hypocreales; Nectriaceae
Diplodia (*Botryodiplodia*) theobromae	Ascomycota; Dothideomycetes; Botryosphaeriales; Botryosphaeriaceae

Erysiphe cruciferarum *Erysiphe heraclei* *Erysiphe polygoni*	Ascomycota; Leotiomycetes; Erysiphales; Erysiphaceae
Fusarium batatis *Fusarium culmorum* *Fusarium oxysporum* *Fusarium oxysporum* f.sp. *basilicum* *Fusarium oxysporum* f.sp. *cepae* *Fusarium oxysporum* f.sp. *cichorii* *Fusarium oxysporum* f.sp. *corianderii* *Fusarium oxysporum* f.sp. cumini *Fusarium oxysporum* f.sp. *vanillae* *Fusarium oxysporum* f.sp. *zingiberi*	Ascomycota; Sordariomycetes; Hypocreales; Nectriaceae
Gloeosporium (*Colletotrichum*) *gloeosporioides*	Ascomycota; Sordariomycetes; Glomerellales; Glomerellaceae
Leveillula (*Oidiopsis*) *taurica*	Ascomycota; Leotiomycetes; Erysiphales; Erysiphaceae
Macrophomina phaseolina Syn. *Rhizoctonia bataticola*	Ascomycota; Dothideomycetes; Botryosphaeriales; Botryosphaeriaceae
Peronospora belbahrii *Peronospora trigonellae* *Peronospora destructor* *Peronospora umbellifarum*	Oomycota; Oomycetes; Peronosporales; Peronosporaceae
Pestalotia cinnamomi	Ascomycota; Sordariomycetes; Amphisphaeriales; Amphisphaeriaceae
Phyllosticta zingiberii *Phyllosticta elettariae* *Phyllosticta murrayae* *Phyllosticta tamarindicola*	Ascomycota; Dothideomycetes; Botryosphaeriales; Botryosphaeriaceae
Phytophthora capsici *Phytophthora nicotianae* var. *nicotianae* *Phytophthora meadii*	Oomycota; Oomycetes; Peronosporales; Peronosporaceae
Plasmodiophora brassicae	Protozoa; Plasmodiophoromycetes; Plasmodiophorales; Plasmodiophoraceae
Protomyces macrosporus	Ascomycota; Taphrinomycetes; Taphrinales; Protomycetaceae
Puccinia allii *Puccinia nakarishikii* *Puccinia porri* *Puccinia psidii*	Basidiomycota; Pucciniomycetes; Pucciniales; Pucciniaceae
Pythium aphanidermatum *Pythium vexans*	Oomycota; Oomycetes; Pythiales; Pythiaceae
Rhizoctonia (Moniliopsis) solani	Basidiomycota; Agaricomycetes; Cantherellales; Certatobasidiaceae

Sclerotinia sclerotiorum	Ascomycota; Leotiomycetes; Helotiales; Sclerotiniaceae
Sclerotium rolfsii	Basidiomycota; Agaricomycetes; Agaricales; Typhulaceae
Septoria lactucae	Ascomycota; Dothideomycetes; Capnodiales; Mycosphaerellaceae
Stemphylium vesicarium	Ascomycota; Dothidiomycetes; Pleosporales; Pleosporaceae
Stromatinia cepivora (Perfect stage of. *Sclerotium cepivorum*)	Ascomycota; Ascomycetes; Helotiales; Scleratiniaceae
Taphrina maculans	Ascomycota; Taphrinomycetes; Taphrinales; Taphrinaceae
Trichoderma	Ascomycota; Sordariomycetes; Hypocreales; Hypocreaceae
Uromyces trigonellae	Basidiomycota; Urediniomycetes; Uredinales; Pucciniaceae
Valsa eugeniae	Ascomycota; Sordariomycetes; Diaporthales; Valsaceae

List of Colour Plates

Glossary

Abaxial – Lower or under surface of leaves.

Acervulus (pl. **Acervuli)** – It is a small, sub epidermal, inverted saucer-shaped asexual fruiting body of some pathogenic fungi that erupts through the epidermis of the host plants. It consists of a mat of hyphae that give rise to short stalked palisade-like conidiophores that cut off conidia from their tips in succession.

Aecium (pl. Aecia) – A cup-shaped fruiting body of the rust fungi consisting of binucleate hyphal cells that produce spore chains consisting of aeciospores following successive conjugate division of the nuclei (Stage I of the rust fungi)

Amphigenous – Fruiting bodies of parasitic fungi occurring on both surfaces of leaves of an infected plant.

Anthracnose – Anthracnose is a kind of fungal spot disease that causes dark, sunken lesions on leaves, stems, flowers and fruits.

Antheridium – The male gametangium of heterogenous fungi.

Apothecium (pl. **Apothecia**) – An open cup-shaped or saucer-shaped ascocarp.

Appressorium (pl. **Appressoria)** – An appessorium is a specialized cell arising from the tip of the germ tube of a germinating fungal spore or a hypha, which is instrumental in infecting the host plant. It is a flattened, thickened hyphal pressing organ that fastens the fungal spore on to the host from which a minute infection peg grows and enters the host by puncturing the cuticle and epidermal cell.

Ascocarp / Ascoma (pl. **Ascomata)** – Fruiting body (Sporocarp) of an Ascomycetous fungus that produces asci and ascospores.

Ascostroma (pl. **Ascostromata)** – Ascostroma is a locule that forms in a stroma, where the asci are formed. This differs from a perithecium that is formed within a stroma in that a perithecial wall is formed by the perithecium that delimits it from the stroma.

Aseptate – Non-septate, coenocytic; lacking cross walls or septa.

Ascus (pl. **Asci)** – A sac-like cell containing a definite number of ascospores, usually eight, formed by free cell formation, mostly after karyogamy and meiosis.

Autoecious / Monoecious – A parasitic fungus that completes its entire life cycle on a single host species as in the case of some rusts.

Basidiocarp – A sexual fruiting body of the basidiomycetous fungi in or on which basidia and basiospores are produced.

Basidium (pl. **Basidia)** – A club-shaped, zygote cell bearing a definite number of basidiospores, usually four, at the end of minute sterigmata formed at the apex of the basidium, following karyogamy and meiosis.

Basipetal – Downward from the apex towards the base. In fungi formation of spores in succession in which the apical spore is the oldest and the youngest spore is at the base.

Blight – A disease characterized by general and rapid killing of tissues of leaves, stems and flowers.

Blotch – A disease characterized by large and irregularly shaped, necrotic spots or blots on leaves, shoots and stems.

Callus – A mass of thin-walled, undifferentiated parenchymatous tissue, usually developed as a result of wounding or infection.

Canker – A necrotic, often sunken or cracked lesion surrounded by callus on a stem, branch or twig of a plant.

Chlamydospores – Thick-walled or double-walled, asexual resting spores formed from hyphal cells, either terminal or intercalary or by the transformation of one or more conidial cells, that can serve as an over wintering reproductive structure.

Chlorosis – Yellowing of normally green tissue due to chlorophyll destruction or chlorophyll formation.

Cleistothecium (pl. **Cleistothecia**) – A completely closed ascocarp containing asci and ascospores.

Coalesce – Grow or join together into a larger spot.

Coenocytic – Non-septate; Aseptate.

Conidiomata – Asexually formed conidia bearing fruit bodies such as pycnidia and acervuli, as well as fruit bearing structures such as sporodochia, synnamata and coremia

Conidiophores – Simple or branched specialized hyphae arising from somatic hyphae, bearing at the tip conidiogenous cell that produces conidia singly or in chains.

Conidium (pl. **Conidia)** – Asexual, non-motile spore formed by abstriction and detachment of part of a hyphal cell at the end of a conidiophore that germinates by a germ tube.

Coremium / Synnema (pl. **Coremia / Synnamata)** – Compact or fused, generally upright conidiophores with branches and terminal spores forming a head-like cluster as in the case of *Fusarium.*

Cystidium (pl. **Cystidia)** – A sterile element occurring in the hymenium of certain Basidiomycetes. They are larger than the other hymenial elements and protrude beyond them.

Demicyclic – A rust fungus that lacks the uredinial (repeating) stage (stage II), but typically has stages 0, I, III and IV.

Dictyospore – A spore with both vertical and horizontal septa.

Dikaryophase – The division of dikaryon having two sexually compatible haploid nuclei simultaneously, leading to the presence of a pair of sexually compatible nuclei in each of the cell.

Diploid – Having two complete sets of chromosomes (2N chromosomes), as a result of fusion of two compatible male and female, haploid nuclei, each having N chromosomes.

Echinulate – Having small, pointed processes or spines projecting from the cell walls.

Epidemiology – Factors influencing the initiation, development and spread of infectious diseases.

Epidermis – Outer layer of cells in plants below the cuticle that forms the external integument in plants.

Erumpent – Bursting or erupting through the substrate surface.

Etiology – The determination and study of the cause or causal organism of a disease.

Facultative parasites – Organisms, which are normally saprophytes, but attain the status of a parasite, when suitable hosts are present and when environmental conditions are favorable to them.

Facultative saprophyte (Facultative saprobes) – Organisms which are normally parasitic on living hosts, start leading a saprophytic life in the absence of suitable hosts.

Falcate – Curved; sickle-shaped.

Flagellum (pl. **Flagella)** – A whip- or hair- or tinsel-like structure that serves to propel a motile cell.

Forma specialis (f.sp.) – A group of races and biotypes of a pathogen species, differing in some physiologic properties, that can infect only plants within a certain host genus or species.

Fructification – Any complex fungal fruiting body that contains or bears spores.

Fusiform – Spindle-shaped, tapering at both the ends as in the case of macroconidia of *Fusarium.*

Gall / Tumor – Abnormal swelling or localized outgrowth, often roughly spherical, but unlike any organ of the normal plant, produced by a plant as a result of attack by a fungus, bacterium, nematode, insect or any other organisms.

Gamete – Sex cell; a reproductive cell, usually haploid that unites with another gamete of the opposite sex to form a zygote.

Geniculate – Bent like a knee.

Germ tube – The initial hyphal growth from a germinating fungal spore, which develops into a mycelium.

Girdle – Encircling and cutting through a stem or the bark and outer few rings of wood and disrupting the phloem and xylem.

Halo – A spot surrounded by a chlorotic zone.

Haploid – A cell or an organism whose nuclei have a single complete set of chromosomes.

Haustoria (sing. **Haustorium)** – Specialized absorbing organs produced by the hyphae, which penetrate the host cells and absorb nourishment from the host.

Heteroecious – Organisms, which require two botanically different hosts for completion of its life cycle, as in the case of some rust fungi.

Honeydew – A sugary exudate secreted by sucking insects, such as aphids, hoppers, mealy bugs, white flies, scale insects etc.

Hyaline – Colorless, transparent.

Hymenium – Hymenium is the tissue layer of the hymenophore of a fungal fruiting body where the cells develop into basidia or asci, which produce spores. In some species all the cells of the hymenium develop into basidia or asci, while in others some cells develop into sterile cells called Cystidia (Basidiomycetes) or paraphyses (Ascomycetes).

Hyperplasia – Abnormal multiplication of cells.

Hypertrophy – Abnormal enlargement of cells.

Hypha (pl. **Hyphae)** – One of the simplest branched filaments of the mycelium of a fungus that is composed of one or more cylindrical cells, which increases in length by growth at its tip, New hyphae arise as lateral branches.

Immune – Having absolute resistance to a particular disease causing pathogen.

Imperfect stage / Anamorph stage – The part of the life cycle of a fungus in which no sexual spores are produced.

Infection peg / Penetrating peg – A fine hypha produced from the underside of an appressorium with a deposition of substances such as, lignin, callose, cellulose, suberin etc around it, that is thrust through the cuticle or epidermis of a host cell to cause infection; Specialized, narrow hyphal strand produced from the underside of an appressorium that penetrates the host cells and causes infection.

Intercalary – Formed along and within the mycelium and not at the hyphal tips.

Intercellular – Hyphae growing in-between the host cells.

Intracellelar – Hyphae penetrating into the host cells.

Isthmus – Pads of gelatinous material formed in-between every pair of sporangia, which function as disjunctor cells that aid in the discharge of the sporangia.

Karyogamy – The fusion of two nuclei of the opposite sexes brought together by plasmogamy into one diploid or zygote nucleus.

Lamina / Leaf blade – The expanded, flat part of a leaf.

Leaf spot – A plant disease lesion typically restricted in development in the leaf after attaining a characteristic size.

Lenticel – A natural opening in the stem of woody plants, bark, tuber, some fruits or root for the exchange of gases between the plant and the atmosphere.

Lesion – Localized, usually well-defined, abnormal change in the structure of an organ or diseased tissue due to disease or injury.

Locule – A cavity within a stroma.

Macrocyclic – A rust fungus that typically exhibits all five stages of the rust life cycle viz., stages 0, I, II, III and IV.

Microcyclic – Rust fungi of genus *Puccinia* that lack pycniospores and aeciospores in their life cycle and always have an autocious life cycle.

Monoecious (syn. Autoecious) – Rust fungi that produce all stages their life cycle on a single plant species; Organisms that produce both male and female reproductive organs in a single individual.

Mosaic – Symptoms of many virus diseases appearing as an abnormal pattern of dark green, light green and chlorotic or yellow areas on leaves.

Motile – Ability of self-propulsion by means of flagella, cilia or amoeboid movement.

Mould – Any profuse or wooly, conspicuous, superficial fungus growth consisting of mycelium and / or spore masses on various substrates, such as surfaces of plant tissues, damp or decaying matter etc.

Mycelium (pl, **Mycelia)** – Mass of hyphae constituting the body (thallus) of a fungus.

Natural openings – Stomata, lenticels, hydathodes and nectarthodes.

Necrosis – Necrosis is a form of disintegration and premature death of cells or tissue; It denotes the death of a circumscribed area of plant tissue accompanied by black or brown darkening due to local toxic or microbiological action as a result of disease or injury.

Obligate parasite – An organism that can live and obtain food only from living protoplasm of the host; It cannot be grown in artificial culture media.

Obligate saprophyte – An organism that can obtain its food only from dead organic matter and is incapable of infecting another living organism.

Oospore – A thick-walled resting spore produced by sexual reproduction.

Ostiole – A pore-like opening in the papilla or neck of a perithecium, pseudothecium, pycnium or pycnidium, through which spores are released.

Papilla – A nipple-like elevation.

Paraphysis (pl. **Paraphyses)** – Sterile, upward growing, basally attached hyphal elements present in the hymenium of the fruiting bodies, such as perithecium, apothecium etc. of some fungi.

Pathogen – A disease-producing organism, agent or factor.

Pedicel – Small, slender stalk, bearing an individual flower, inflorescence or spore.

Pediculate – Not sessile; Having a pedicel.

Penetration peg – The specialized, narrow, hyphal strand produced from the underside of an appressorium that penetrates the host cells.

Perfect stage – The sexual reproductory stage in the life cycle of a fungus.

Periphyses (sing. **Periphysis)** – Short, thread-like filaments that line the opening or ostiole of a perithecium in certain types of fungi.

Perithecium (pl. **Perithecia)** – A flask-shaped or subglobose, thin-walled sexual fruiting body of a fungus (ascocarp), containing asci and ascospores, which are expelled or released through a pore (ostiole) at the apex.

Phialide – A type of bottle-shaped conidiogenous cell that produces blastic conidia in basipetal succession.

Planogamete – A motile gamete.

Plsmodium – A naked, multinucleate mass of protoplasm moving and feeding in amoeboid fashion.

Plasmogamy – Union of protoplasts of two cells that brings the nuclei of the two cells close together within a single cell.

Propagule – Any unit or part of an organism, such as spore, chlamydospore, sclerotium, mycelial fragment etc. capable of independent growth to produce a new individual.

Pseudothecium (pl. **Pseudothecia**) – An ascocarp similar to a perithecium, but unlike typical perithecium produces asci in unwalled locules.

Pustule – A blister-like or pimple-like, eruptive fruiting structure, such as uredinium of a rust fungus.

Pulvinate – Having a swelling at the base.

Pycnidium (pl. **Pycnidia**) - An asexual, thick-walled, spherical or flask-shaped fruiting body, lined inside with conidiophores that produce pycnidiospores (conidia) from their tips.

Pycnium / Spermagonium (pl. **Pycnia / Spermagonia**) – Globose or flask-shaped, haploid fruiting body of rust fungi bearing pycniospores / spermatia and receptive hyphae.

Rachis – Elongated main axis of an inflorescence.

Resting spore – A sexual or asexual, thick-walled spore of a pathogen that is resistant to extremes of temperature and moisture and which often germinates only after a resting period of dormancy from its formation.

Rhizome – A mostly horizontal, jointed, fleshy, often elongated, usually underground stem with nodes, buds and scales.

Sclerotium (pl. **Sclerotia)** – A hard resting body, resistant to adverse environmental conditions and may remain dormant for long period of time and germinates when favorable conditions return; it consists of closely packed, more or less isodiametric or oval cells, known as pseudoparenchyma tissue.

Seta (pl. **Setae)** – Dark-colored, stiff, bristle-like or hair-like, sterile structure occurring in the fruiting body (acervuli) of some fungi.

Shot holes – A symptom caused by some leaf spotting pathogens in which necrotic areas of limited size fall out from the lesions on the leaf lamina, leaving small, almost circular holes, which appear as holes made by bullet shots.

Soft rot – A rot of fleshy fruit, vegetable or ornamental in which the tissue becomes macerated by the enzymes of the pathogen.

Sooty mould – A black, sooty coating on the foliage and fruits formed by the dark hyphae of the fungi that live on the honey dew secreted by sucking insect pests, such as aphids, plant hoppers, mealy bugs, scale insects etc.

Sporocarp – spore bearing fruiting body

Sporodochium (pl, **Sporodochia)** – An erumpent crowded cluster of conidiophores without any laeral union arising from a stroma in the form of a cushion as in Tuberculariaceae

Sterigma (pl. **Sterigmata)** – A small, tender, pointed projection that supports a spore.

Stoma (pl. **Stomata)** – A structure composed of two guard cells and an opening in-between them in the epidermis of a leaf or stem.

Stroma (pl. **Stromata**) – A compact somatic structure, much like mattress or cushion made up of loosely woven hyphae, where fructifications are usually formed.

Susceptible – Capable of being infected; lacking the inherent ability to resist disease or attack by a pathogen; not immune.

Symptom – A visible, abnormal change in a host and its behavior as a result of infection by a pathogen.

Target spot – A lesion consisting of a dark brown circular area containing a series of brown, concentric rings appearing like a target board; typical of infection caused by species of *Alternaria.*

Teliospore / Teleutospore – A thick-walled resting spore of the rusts and smuts in which karyogamy occurs.

Telium (pl. **Telia**) – The fruiting body (sorus) of a rust fungus that produces teliospores.

Thallus – Vegetative body of fungus. that is devoid of stem, roots and leaves or soma of fungi.

Toxin – An organic poisonous substance produced by a microorganism, which even at low concentrations deleteriously and irreversibly affects the normal physiological processes of living organisms.

Transpiration – The evaporation of moisture from a living plant, mainly through the stomata of the leaves.

Tumor – An uncontrolled overgrowth of tissue or tissues.

Uredinium / Uredium – The sorus of rust fungi producing binucleate, repeating urediniospores / uredospores.

Vascular wilt disease – A disease in which the pathogen is almost entirely confined to the vascular system of the host during pathogenesis and in which wilting is a characteristic symptom.

Virulent – Capable of causing a severe disease; strongly pathogenic

Virus – An ultramicroscopic, filterable, intracellular, infective agent of obligate nature, consisting of a core of infectious nucleic acid, either RNA or DNA, usually surrounded by protein coat and capable of causing various diseases in living organisms.

Volunteer plant - A plant from a previous season's crop that regenerates in a subsequent crop or plant developing from a self-sown or lost seed.

Water-soaked – A disease symptom, in which an area of plant cells become darker in color owing to the filling of intercellular air spaces with cell sap, which comes out to the surface as exudates.

Wet rot – Any kind of rot, in which the tissue is rapidly and completely disintegrated with release of water from the lysed cells.

White rot – Rotting of wood in trees invaded by lignin destroying fungi that leave a white cellulose residue.

Wilt – Loss of turgidity and drooping of plant parts generally caused by insufficient water or excessive transpiration in the plant; a disease symptom caused by the wilt fungi.

Wound parasite – A parasitic organism, which can invade a host only if it can first become established in damaged tissues.

Xylem – It is a conducting tissue consisting of tracheids, vessels, parenchyma cells and fibers, which helps in conduction of water and minerals from the roots to the different parts of the plant.

Yellows – A plant disease characterized by pronounced chlorosis, yellowing and stunting of the host plant as a result of infection by a pathogenic organism.

Zoospores – Non-sexual, motile sporangiospores or swarm spores produced from either sporangia or zoosporangia, having one or two flagella.

Zygosporangium – A sporangium that arises from a zygospore at the end of a stalk, which contains non-motile spores.

Zygospore – The zygote formed as a result of fusion of two morphologically identical gametangia.

Zygote – A diploid cell formed by the union of two, compatible gametes, as well as the individual produced from such a cell.

1

Introduction

A **'spice'** is a seed, fruit, root, bark, bulb, corm, rhizome or any other part of a plant primarily used for enhancing the taste, flavor or color of food. Some spices serve as preservative of food substances. Many spices have antimicrobial properties. Spices are sometimes used in medicines, religious rituals, cosmetics, perfumes etc. Spices are more commonly used in countries having warmer climates, especially in meat preparations. Many spices have substantial antioxidant property due to the presence of phenol compounds, particularly flavanoids.

India is the major producer of spices and contributes about 75 per cent of world's spice production. Developing and under developed countries of the world supply about 55 per cent of spices to the global markets and the major markets in global spice trade are the United States of America, The European Union, Japan, Singapore, Saudi Arabia and Malaysia.

Top spice producing countries of the world and production (in metric tons)

Country	2010	2011
India	14,74,900	15,25,000
Bangladesh	1,28,517	1,39,775
Turkey	1,07,000	1,13,783
China	90,000	95,890
Pakistan	53,647	53,620
Iran	18,028	21,307
Nepal	20,360	20,905
Columbia	16,998	19,378
Ethiopia	27,122	17,905
Sri Lanka	8,293	8,428

Source - UN Food and Agriculture Organization.

According to the International Standards Organization, there are about 109 species of spice crops grown in different parts of the world under varied climatic conditions and about 63 species are cultivated in India.

Important spice crops

- Ginger (*Zingiber officinale*)
- Black pepper (*Piper nigrum*)
- Cardamom (*Elettaria cardamomum*)
- Clove (*Syzygium aromaticum*)
- Turmeric (*Curcuma longa*)
- Garlic (*Allium sativum*)
- Coriander (*Coriandrum sativum*)
- Fenugreek (*Trigonella foenum - graecum*)
- Cumin (*Cuminum cyminum*)
- Black cumin (*Nigella sativa*)
- Aniseed/Sweet cumin (*Pimpinella anisum*)
- Mustard (*Brassica juncea*)
- Black mustard (*Brassica nigra*)
- White mustard (*Sinapis alba*)
- Cinnamon (*Cinnamomum verum / C. zeylanicum*)
- Cassia (*Cinnamomum cassia / C. aromaticum*)
- Allspice (*Pimenta dioica*)
- Nutmeg (*Myristica fragrans*)
- Vanilla (*Vanilla planifolia / V. tahitensis*)
- Basil (*Ocimum basilicum*)
- Fennel (*Foeniculum vulgare*)
- Saffron (*Crocus sativus*)
- Chicory (*Cichorium intybus*)
- Curry leaf (*Murraya koenigii*)
- Lemon grass (*Cymbopogon citratus*)
- Tamarind (*Tamarindus indica*)
- Red chilli / Paprika / Cayenne pepper (*Capsicum annuum*)
- Green chilli / Chilli pepper (*Capsicum* spp.)
- Asafoetida (*Ferula asafoetida*)

2

Diseases of Spice Crops

1. Ginger *(Zingiber officinale)*

Ginger Plants

Ginger rhizomes

'Ginger', which originated in Southeast Asia has a long history that goes back over 5000 years when the Indians and Ancient Chinese considered it as a tonic root for all ailments. At an early date it was exported to Ancient Rome from India and was used extensively by the Romans. Subsequently it was cultivated in several other countries for its medicinal value. Ginger is cultivated in most of the states of India. Kerala and Meghalaya are the major ginger growing states. India's production of ginger constitutes about 50 per cent of the total world's production.

Ginger is an erect, herbaceous , perennial, flowering plant, the rhizome of which is used as a spice. Ginger is an ancient spice and is a common ingredient in many Asian foods and is used in soups, curries, noodles, stews etc. It is widely used to flavor beverages, such as tea, coffee, lemonades, cocktails etc. In many places across the world ginger is used in sweets, alcoholic beverages, such as ginger beer and wine. Ginger has powerful anti-inflammatory properties. Further it has many other health related benefits, such as ability to help with arthritis, osteoarthritis, relieves nausea and pain, prevents cancer, improves respiratory conditions and reduces flatulence. It also boosts bone health, strengthens the immune system and increases appetite. Certain chemical compounds in fresh ginger help in warding off germs and are especially good at halting growth of bacteria, such as *Escherichia coli* and *Shigella,* and may also keep viruses like 'Respiratory syncytial virus' (RSV), which causes infections of the lungs and respiratory tract at bay. Ginger also enhances sexual activity, mitigates obesity and relieves pain related menstrual disorders.

Diseases of ginger (*Zingiber officinale*)

i) Soft rot / Rhizome rot of ginger - *Pythium aphanidermatum*

'Soft rot of ginger' is the most serious and destructive disease of ginger and causes rotting of the rhizomes both in the field and even after harvest. Due to this disease more than 50 per cent yield loss has been reported. In severe cases the entire clump rots completely leading to total yield loss. The disease is mainly caused by *Pythium aphanidermatum,* however *P. myriotylum, p. vexans* and a few other species are also associated with the disease Infection starts at the collar region of the pseudostem and progresses both upwards and downwards. Infected pseudostem becomes water soaked and begins to rot. The rotting gradually spreads to the rhizome leading to soft rot. The rotten pulpy mass of tissues are covered by a fairly thick rind. The rotten rhizomes emit a foul odor. Ultimately the disease spreads to the roots and they also rot. Foliar symptoms appear as light yellowing of the tips of lower leaves and spread to the leaf blades. In the beginning the middle portion of the leaves remain green while the margins turn yellow. Ultimately the yellowing spreads to all the leaves from the bottom upwards resulting in drooping, withering and drying of the pseudostem. The affected pseudostem comes off easily when pulled slightly. Younger sprouts are highly susceptible to infection by the disease.

Rhizome rot of ginger

The pathogen is both seed- and soil-borne. The inoculum present in the seed rhizome initiate the disease in the field. The resting bodies, such as oospores and chlamydospores, which get access to the soil from infected rhizomes survive in the field for prolonged periods and cause initial infection. Secondary spread is through zoospores produced from the sporangia, which are carried to other fields by cattle, agricultural implements and even by man.

High temperatures ranging from 28° - 35°C, high soil moisture, water logging, poor drainage and presence of abundant decomposing organic matter in the soil predispose the crop to infection and rapid spread of the disease.

Control measures

- Use of healthy rhizomes as seed material is very effective in controlling the disease.

- Crop rotation with non-host crops, such as legumes, maize, ragi and paddy helps to keep the disease under check.
- Diseased clumps have to be removed and burnt. Steeping the rhizome seed pieces in hot water at 50ºC for 10 minutes before sowing affords protection from disease occurrence.
- Steeping the rhizome seed pieces in metalaxyl (Ridomil MZ or Apron) 0.2% or thiophanate-methyl (Topsin M) 0.2% or fosetyl-Al (Aliette) 0.1%) or chlorothalonil (Daconil) 0.15%) and air drying the same prior to sowing affords effective protection from disease occurrence.
- Rhizome seed treatment with *Trichoderma harzianum* at 5.0 g / Kg of rhizome seed material prior to sowing is also effective .
- Because high soil moisture and water logging are favorable for the disease occurrence and spread, proper drainage facilities should be provided. Cultivating the crop in raised beds of 30 cm height and 1.0 m width is recommended.
- Soil amendment by application of powdered oilseed cakes of *Azadirachta indica*, *Callophyllum inophyllum*, *Pongamia glabra*, *Hibiscus sabdariffa* and *Brassica campestris* at 150 - 200 Kg / ha helps to minimize the incidence of the disease.
- Drenching the seed pieces in cow dung slurry mixed with *Trichoderma harzianum* at 10 g / litre of water before sowing has been found to be effective.
- Drenching the seed pieces with Bordeaux mixture 1.0% or Mancozeb 0.2% or difolaton 0.15% at the time of sowing and again after germination at 2 - 3 weeks interval gives partial control.
- Treating the seed rhizomes in aerated steam at 40ºC for 60 minutes eliminates both externally and internally seed-borne inoculum.

For seed purposes, disease-free, well matured, healthy rhizomes are selected and stored in sand or pits of suitable size (1.0 x 2.0 m^2) under shade. Before storage the seed rhizomes are treated with a mixture of Quinalphos 0.075% and Mancozeb 0.3% for 30 minutes and with *Trichoderma harzianum* for 30 minutes and shade dried. The lateral walls of the pit are plastered with cow dung paste. Sand or sawdust is spread to a height of 5.0 cm at the bottom. Over this a layer of treated seed rhizomes are arranged and then covered with a layer of sand or saw dust to a height of 2.0 cm. Thus several layers of rhizomes are arranged alternating with sand or saw dust. Some space is left empty at the top to provide aeration. Then the pit is covered with a wooden plank and sealed with clay.

Thus the seed rhizomes are stored for a period of 3 - 4 months and then used for planting during the ensuing season. Cultivars, such as Maran, Nadiya and Narasappatom are reported to be resistant to the disease.

General characters of genus *Pythium* - Refer page - 133

ii) *Phyllosticta* leaf spot of ginger - *Phyllosticta zingiberi*

The symptoms of this foliar disease appear as small, scattered, oval to elongated, water-soaked spots in large numbers on the leaves. The spots are distributed irregularly over the entire surface of leaves. The spots turn whitish and papery with dark brown margins with a yellow halo. The pathogen forms amphigenous, pinpoint-like, sub-globose, dark brown, ostiolate pycnidia on the upper surface of the spots. Pycnidiospores are released in large numbers from the pycnidia through an ostiole. The fungus over winters in infected plant debris and seed rhizomes. It can survive in the diseased plant debris in the soil for more than 12 months and continue to produce pycnidiospores.

Phyllosticta leaf spot of ginger

Temperature ranging from 20° - 30°C and relative humidity 80 - 90% are found to be favorable for disease occurrence and spread. The disease incidence is more severe during the monsoon periods. Heavy rainfall accompanied by strong wind results in splashing of the pathogen to greater distances thereby causes increased disease incidence.

Control measures

- Infected plant debris to be collected and burnt. Healthy rhizomes collected from disease free plants have to be used as seed material
- Before sowing, the seed rhizomes are steeped in mancozeb 0.25%, copper oxychloride 0.25% or carbendazim 0.1% solution for 60 minutes and shade dried.
- At the onset of the disease the crop has to be sprayed with mancozeb 0.2% or carbendazim 0.1% or a combination of the two fungicides. Fungicidal application has to be continued at 10 - 14 days interval depending upon the intensity of the disease.
- Cultivars, such as Narasapatom, Tura, Nadia, Tetraploid and Thingpani have been reported to be moderately resistant to the disease.

General characters of genus *Phyllosticta* - Refer page - 135

iii) *Colletotrichum* leaf spot of ginger - *Colletotrichum zingiberis*

Initial symptoms appear as small, circular to oval, pale yellow spots. The spots enlarge along the length of the leaves and turn brown. The spots may coalesce to form irregular, brown patches. Later the central portion of the spots become necrotic, develop cracks, dry and fall off leaving ragged holes. Severely affected leaves turn yellow completely, dry and die. Linear, brown lesions are also formed on the leaf sheaths and scales. Minute, dark-colored. pin point-like dots representing the acervuli are formed on the spots in an irregularly concentric pattern.

The pathogen survives in the diseased plant debris in the soil for prolonged periods and continues to produce conidia, which are carried by air and cause primary infection. Secondary infection occurs through air-borne conidia produced from the acervuli on the plants.

The disease occurrence and spread is more during the monsoon seasons when there is continuous rain for several days. Moist and damp weather conditions and moderate temperatures favor disease occurrence and spread.

Control measures

Diseased crop debris in the field should be removed and destroyed. Spraying the crop with Bordeaux mixture 1.0% or copper oxychloride 0.25% or mancozeb 0.2% when initial symptoms of the disease is noticed followed by 2 - 3 sprayings at fortnightly intervals has been found to be effective in controlling the disease.

General characters of genus *Colletotrichum* - Refer page - 136

iv) Yellow disease of ginger/*Fusarium* yellows of ginger - *Fusarium oxysporum* f.sp. *zingiberis*

Yellow disease of ginger

The pathogen is specific to ginger. Infection occurs as a fast developing rot, when the pathogen is carried in the seed rhizome, as a result germination is affected or the shoots that emerge are very week and they soon die. If the seedlings survive this early rotting or when infection occurs in the seed rhizomes later on through injuries or wounds, symptoms appear rather slowly. Such affected plants become stunted and the leaves turn yellow starting from the lower leaves, which dry and die. Gradually the disease progresses to the upper leaves, which show yellowing and finally leading to the premature death of all above ground parts. The rhizomes below the ground shrivel and develop a brown or black rot of the outer layers. The affected rhizomes do not develop water-soaked rotting as in the case of bacterial soft rot, but show symptoms of dry rot. The water conducting vessels become brownish. Finally only the shell of the rhizome and fibers are left over. A white, cottony growth comprising of fungal mycelium and spores of the pathogen develop on the rhizome.

The disease is both soil-borne and seed-borne. The sclerotia and even the mycelium of the pathogen can remain in the plant debris and soil for prolonged periods and perpetuate the disease. The pathogen present in the infected seed rhizomes also causes initial infection. Secondary spread may take place through blowing wind, irrigation water, implements or even by man. Moderate temperatures, high relative humidity, heavy rainfall, poor drainage facilities etc. predispose the crop to infection and spread of the disease.

Control measures

Avoiding cultivation of ginger in sick soils, adopting crop rotation practices, removal and destruction of crop debris from the field, use of disease-free, healthy seed rhizomes for seed purposes, removal and destruction of disease affected clumps, providing proper drainage facilities, avoiding injuries or wounds to rhizomes during cultural operations and while harvesting, transit , and storage etc. help to minimize the incidence of the disease.

Steeping the rhizome seed pieces in Captan 0.2% or Thiram 0.2% solution for 60 minutes and shade drying before sowing helps to eradicate externally seed-borne inoculum. Drenching the sowing furrows with propiconazole 0.1% or carbendazim 0.2% solution immediately after sowing helps to control initial infection.

Soil application of the biocontrol agents viz., *Trichoderma viride, Trichoderma harzianum or Pseudomonas fluorescens* inhibits the growth and multiplication of the pathogen.

General characters of genus *Fusarium* - Refer page - 138

v) Sheath blight / Leaf blight of ginger - *Rhizoctonia (Moniliopsis) solani*

Sheath blight of ginger

The disease is considered to be a minor one and occurs occasionally. Symptoms of the disease appear as small lesions on the leaf sheaths, but may develop on the leaf blades also. Initial lesions are small, ellipsoid or ovoid and greenish-gray in color. Under favorable conditions the lesions enlarge and coalesce to form bigger lesions with irregular border and with grayish-white centre, and dark brown border. Development of several such large spots on the leaf sheath causes the death of the whole leaf. Numerous, pinhead-like, black sclerotia are formed on the affected leaf sheaths and base of the plant.

The pathogen is both soil- and seed-borne. The fungus survives in the plant debris in the soil for many years as sclerotia and causes primary infection. The pathogen is also carried through infected rhizomes.

Warm temperatures ranging from 15° - 18°C and wet weather conditions favor the disease occurrence and spread.

Control measures

Deep summer ploughing, use of disease-free rhizomes as seed material, adopting crop rotation practices, providing adequate drainage facilities, application of farm yard manure or neem cake to encourage growth of antagonistic microorganisms, removal and destruction of plant debris in the field and other such field sanitation measures help to reduce the inoculum and occurrence of the disease.

Treating the rhizome seed pieces in hot water at 47°C for 30 minutes helps to eradicate the seed-borne inoculum.

Steeping the rhizome seed pieces in mancozeb 0.2% solution for 60 minutes and shade drying before sowing is effective in eradicating externally seed-borne inoculum.

General characters of genus *Rhizoctonia* - Refer page - 140

vi) Dry rot of ginger - *Fusarium oxysporum* f.sp. *zingiberis* and *Pratylenchus coffeae*

The disease mostly appears late in the growing season of the crop. The affected plants appear stunted and exhibit varying degrees of yellowing. The symptoms appear first in the older leaves and then spread to all the leaves. The affected rhizomes and roots show brownish lesions, which later merge and cover large areas. In advanced stages, the rhizomes when cut show a brownish ring, which is mainly restricted to the cortical region.

Dry rot of ginger

The pseudostem of dry rot affected plants do not come off easily with a gentle pull in contrast to soft rot affected plants. The surface of affected rhizomes are discolored and the rhizomes are often shrunken and dry. The fungus and the nematodes are found in the rotten rhizomes.

Soil and the infected rhizome seed pieces are the sources of primary infection. The fungus produces large number of chlamydospores, which are found in the decomposing tissues of infected rhizomes and perpetuate the disease.

Warm, wet weather conditions and high soil moisture favor disease occurrence and spread. The pathogen can get easy access to the roots and rhizomes of ginger plants through wounds caused by the nematodes.

Control measures

- Deep summer ploughing, use of disease-free rhizomes as seed material, adopting crop rotation practices, providing adequate drainage facilities, application of farm yard manure or neem cake to encourage growth of antagonistic microorganisms, removal and destruction of plant debris in the field and other such field sanitation measures help to reduce the inoculum and occurrence of the disease.
- Treating the rhizome seed pieces in hot water at 47°C for 30 minutes helps to eradicate the seed-borne inoculum.
- Steeping the rhizome seed pieces in mancozeb 0.2% solution for 60 minutes and shade drying before sowing is effective in eradicating externally seed-borne inoculum.

General characters of genus *Fusarium* - Refer page - 138

vii) Bacterial wilt of ginger - *Ralstonia solanacearum*

'Bacterial wilt of ginger' is another serious and destructive disease of ginger and causes extensive damage to the crop and yield loss.

Bacterial wilt of ginger

Initial symptoms appear as small, water-soaked spots at the collar region of the pseudostem, which spread in all directions, as well as upwards and downwards. This is followed by mild drooping and curling of the leaf margins of the lower leaves, which spread upwards to all the leaves. Yellowing starts from the lowermost leaves and progresses gradually to the upper leaves. In advanced stages severe yellowing and wilting symptoms occur. The pseudostem of such affected plant comes off easily when pulled slightly. The vascular tissues of the affected pseudostems show dark streaks. The affected pseudostem and rhizome when pressed gently exude a milky ooze from the vascular strands. Finally the rhizomes get rotted completely and the plants die.

Bacterial wilt is a soil and seed-borne disease. High soil moisture above 50% and high temperature of 37°C favor disease occurrence and spread. The disease occurs mostly during the south west monsoon season.

The bacteria spread through soil, water and infected rhizomes. The bacteria get access to the plants through wounds or aberrations caused in the rhizomes and roots at the time of planting, cultural operations, nematodes and insects.

Control measures

- Use good quality, disease-free seed rhizomes for sowing
- Avoid water-logging in the field by providing suitable drainage facilities.
- Remove disease affected clumps from the field and destroy them by burning.
- Keep the field clean from weeds.
- Crop rotation with non-host crops, such as legumes, paddy, ragi, maize etc. is effective in preventing occurrence of the disease.
- Treat seed rhizomes with Streptocycline 200 ppm for 30 minutes and shade dry before sowing.
- Hot water treatment of the seed rhizomes at 50°C for 10 minutes has been found to afford protection from the disease.
- Once the disease appears in the field drench the soil especially around the collar region with Bordeaux mixture 1% or copper oxychloride 0.25%

viii) Storage rot of ginger - *Fusarium oxysporum* f.sp. *zingiberis* and *Sclerotium rolfsii*

'Storage rot of ginger' is a post-harvest disease that affects the ginger rhizomes during storage. Infection of the rhizomes may take place in the field before harvest or at the time of harvest or during handling or transit through injuries. Several fungal and bacterial microorganisms may be responsible for causing storage rot. However, *Fusarium oxysporum* f.sp. *zingiberis* and *Sclerotium rolfsii* are the two most important organisms responsible for causing storage rot. In the case of infection by *Fusarium* sp., the rot occurs mostly as soft rot and on the rotten areas whitish, fine, cottony growth comprising of the fungal mycelium and spores develop. In the case of attack by *Sclerotium* sp., minute, black sclerotia are formed on the infected area. Under favorable conditions for the pathogens, the rotting may cover the entire rhizome.

The disease is carried through infected rhizomes and secondary infection may take place in the store house from the infected rhizomes. Warm and humid conditions favor the disease occurrence and spread.

Control measures

- Harvesting should be taken up on bright sunny days.
- Care should be taken to avoid injuries or wounds to rhizomes at the time of harvest, handling and transit.
- The harvested rhizomes should be properly dried before storage.
- Disease affected and damaged rhizomes should be sorted out and removed from the bulk consignment before storage.
- The rhizomes should be stored in clean store houses with good aeration.

General characters of genus *Sclerotium* - Refer page - 141

2. Black pepper *(Piper nigrum)*

'Black pepper' is a perennial, woody, flowering vine belonging to the family Piperaceae that grows on supporting trees, poles or trellises. It is cultivated for its fruits known as 'pepper corns'. A single stem may bear 20 - 30 fruiting spikes.

Black pepper is native to present day Kerala in South India. It is extensively cultivated in Kerala, Karnataka and Tamil Nadu as well as in many tropical countries, such as Vietnam, Indonesia, India and Brazil. Vietnam is the world's largest producer of pepper, producing 39 per cent of the world's ginger as on 2016.

Pepper Plants

Pepper Corns

Pepper Corns

Black pepper is one of the most commonly used spices worldwide and is deemed to be the **'King of spices'**. It is used widely as a spice in several vegetarian and non-vegetarian dishes, as well as a medicine. The chemical 'piperine' present in black pepper causes spiciness. The antioxidants present in black pepper help to prevent many diseases. The antioxidants aid in weight loss and help in relieving sinus, asthma, and nasal congestion. They also reduce the risk of cancer and heart, and liver ailments.

Diseases of black pepper (*Piper nigrum*)

i) Foot rot / Quick wilt of pepper - *Phytophthora capsici*

'Foot rot' or **'Quick wilt'** of pepper, otherwise called as **'Berry spot'** or **'Berry split'** is a severe and destructive disease of black pepper and all cultivated varieties of black pepper are susceptible to the disease.

Quick wilt / Foot rot of pepper

Initially water-soaked spots are formed on the collar region of the plant, which change into wet, slimy dark patches and rot, as a result the affected plant wilts rapidly. This is known as the **'foot rot'** phase and is the most destructive phase of the disease Water-soaked dark patches appear on the roots resulting in root decay and shredding of roots. Infection spreads to the runner shoots and the sprouts also rot resulting in shoot rot. The rotting extends both upwards and downwards and the branches wilt and die from the tip downwards. This is referred to as the **'die back'** phase of the disease. The rotten tissues of the affected regions emit a foul odor. The stems break at the nodal regions and the branches turn dark brown due to rotting. Usually the infection starts from the feeder roots and then spreads to the lateral roots and finally to the main roots. When the root system is affected, the plant wilts suddenly even before collar infection. This is referred to as **'quick wilt'.** On the leaves brownish, water-soaked, concentric spots with fimbriate margins appear. The central portion of the spots turn grayish. These lesions rapidly enlarge into large, dark brown patches covering large areas of the leaves. The under surface of the spots may be covered with a delicate, downy growth of the pathogen. Severely affected leaves turn yellow, wilt and drop off prematurely. Small, oval, dark brown to black, slightly sunken spots develop on the spikes and berries. The berries shrivel and wither followed by heavy spike shedding. This is known as **'berry spot'**.

The oospores and chlamydospores surviving in the soil initiate infection in the roots and collar region. Secondary infection is through zoospores emerging from the oospores and sporangia. Pepper being a perennial plant, the pathogen may continue to survive in the live plant tissues and spread the disease all through the year. Further the pathogen has many collateral and alternate hosts and so there is a continuous supply of spores from such sources, which may perpetuate the disease.

Heavy rainfall, more number of rainy days, high relative humidity of more than 90%, low temperature range of 22° - 25°C and less sun shine hours of 2.8 - 3.5 hours per day predispose the plants to disease occurrence and spread.

Control measures

- Infected plant debris and dead vines along with the entire root system are removed and destroyed by burning, so as to reduce build up of inoculum. Drench such spots with Bordeaux mixture 1.0% or Copper oxychloride (0.25%)
- Use disease free planting material.
- Provide adequate shade and follow proper pruning technique.
- Apply 1.0 Kg of lime and 2.0 Kg of neem cake per standard per year before the monsoon rains.
- Treat the pepper vine transplants with *Trichoderma harzianum* or *T. viride* or *Pseudomonas fluorescens* at the time of planting.
- Soil drenching around the stem of vines with Bordeaux mixture 1.0% or copper oxychloride 0.25% at the rate of 3 - 4 litres per vine, 4 times at monthly intervals and foliar spraying with Bordeaux mixture 1.0% or copper oxychloride 0.25% , 8 times at fortnightly intervals starting from the onset of the south west monsoon affords good control of the disease.
- Soil drenching with metalaxyl 0.1% or chlorothalonil 0.125% is also effective in controlling the disease
- Foliar spraying with metalaxyl 0.1% or fosetyl Al. 0.1% , 3 times at monthly intervals has also been found to afford good control of the disease.

General characters of genus *Phytophthora* - Refer page - 142

ii) 'Pollu disease' / Anthracnose of pepper - *Colletotrichum gloeosporioides* and *C. necator*

'Pollu' disease or **'Anthracnose'** is a severe disease of black pepper. Symptoms appear on the leaves as small, circular or angular brown spots with a prominent yellow halo. Similar spots appear on the stem portions also, which spread downwards. Spots on leaves coalesce resulting in leaf blight. On the stem, the pathogen infects the tip region and black, sunken, linear necrotic lesions appear, which may extend downwards. The lesions may spread and girdle the affected area and the branch starts drying from the tip downwards. On the dried, dead areas, black colored acervuli are formed, which appear as minute pin-point like dots. When the spikes are affected black discoloration appears at the point of attachment of the spikes with the stem and the spikes are shed leading to heavy yield loss. Sunken lesions develop on the infected berries and the berries get mummified and become hollow, which is termed as **'pollu'**. Radially split cracks develop on the pericarp of berries. Concentric rings of acervuli are formed on the upper surface of infected leaves also.

Primary infection is caused by sowing infected seeds. Secondary infection occurs through conidiospores and ascospores spread by blowing wind and rain splash.

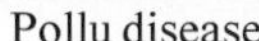
Pollu disease

Anthracnose of pepper

The pathogen has many other alternate hosts and so there is a steady supply of spores, which are carried to pepper crop through wind, rain and mechanically by some insects.

Moderate temperatures of about 25°C, high humidity of 95 - 97% and heavy rainfall favor the occurrence and spread of the disease.

Control measures

- Removal and destruction of fallen leaves and spikes help to eliminate the inoculum.

- Providing adequate shade and irrigation during the summer months and application of recommended doses of fertilizers help to mitigate the disease incidence.
- Spraying the crop with Bordeaux mixture 1.0% or Copper oxychloride 0.25% or Carbendazim 0.1% or Mancozeb 0.2% or Thiophanate methyl 0.1% or Chlorothalonil 0.125% gives effective control of the disease. Spraying should be taken up as soon as disease symptoms are noticed, followed by 2 - 3 sprayings at fortnightly intervals.

General characters of genus *Colletotrichum* - Refer page - 136

iii) Charcoal rot / Basal rot of pepper - *Rhizoctonia bataticola* (Synonym *Macrophomina phaseolina*)

The disease is primarily soil-borne. The microsclerotia produced by the fungus in large numbers, which remain in the soil and infected crop residues in a viable state for a considerably long period of time initiate infection. The pathogen has a wide host range including groundnut, tobacco, beans, soybean, pigeon pea, sorghum, corn etc.

Symptoms appear as discoloration of the vine at the soil level followed by formation of cankers on the stem region, which may spread upwards. The leaves wilt and drop from the plant. Numerous small, black sclerotia are produced by the fungus in the affected tissues. The stem may develop cracks.

Relatively high temperature and drought conditions favor disease occurrence and spread.

Control measures

- Three year crop rotation with non-host crops is helpful in minimizing the inoculum potential in the soil.
- Application of neem seed cake to the soil helps to reduce the inoculum in the soil.

General characters of genus *Rhizoctonia* - Refer page - 140

iv) Slow decline / Slow wilt of pepper - *Phytophthora capsici, Radopholus similis* and *Meloidogyne incognita*

'Slow wilt' of pepper is caused by the association of a few fungal and nematode organisms. However, *Phytophthora capsici, Radopholus similis* and *Meloidogyne incognita* are mainly responsible for causing the disease. Symptoms of the disease appear as general yellowing of the lower leaves, which progress upwards and cover all the leaves. Affected leaves become flaccid and fall off prematurely. Tip burning of leaves and die-back of twigs appear. Affected vines die gradually. Vascular browning of the stem is also seen. The root system of affected vines show varying degrees of necrosis. Root knots can also be seen in vines showing yellowing of leaves. Root galls appear due to nematode infestation leading to rotting of feeder roots. Both the fungal pathogen and the nematodes are responsible for causing damage to the feeder roots.

Slow wilt of pepper

The causal organisms are mainly soil borne. The fungus can survive in the plant debris and soil for long periods and perpetuate the disease. Cysts and egg masses of the nematodes also remain in the infected plant debris and soil for prolonged periods. The pathogens may also be dispersed through running water. Rainy seasons and light loamy soils favor disease occurrence and spread.

Control measures

- Severely affected vines may be removed and destroyed.
- Soil application of neem seed cake at 2.0 Kg / vine helps to eliminate the fungus and nematode pests by encouraging the growth of antagonistic microorganisms in the soil
- Drenching the soil with copper oxychloride 0.25% controls the fungal pathogen.
- Soil application of phorate 10G at 80g / vine or carbofuran 3G at 100g / vine is effective in controlling the nematodes.

General characters of genus *Phytophthora* - Refer page - 142

v) Leaf rot and Blight of pepper - *Rhizoctonia solani / Moniliopsis solani*

The disease is often serious in nurseries and the pathogen attacks the seedlings more than grown up plants. The fungus infects both leaves and stems. Grayish, sunken lesions and mycelial threads develop on the leaves and the infected leaves are attached to one another with the mycelial threads. On the stems the symptoms appear as dark brown lesions, which spread both upwards and downwards. The new flushes arising from the points of infection gradually wilt, droop and dry up. On the mature lesions black, pinhead-like sclerotia are formed in large numbers.

The pathogen is soil-borne and the sclerotia can remain in a viable state in crop debris and soil for a prolonged period and cause primary infection.

The disease becomes more severe when warm and humid weather conditions prevail. High soil moisture favor rapid spread of the disease.

Control measures

Removal and destruction of plant debris from the soil, providing adequate drainage facilities, removal and destruction of severely affected plants, eradicating alternate and collateral hosts of the pathogen are some of the methods that can be followed to prevent the occurrence of the disease.

Drenching the cuttings after planting with Bordeaux mixture 1.0% or copper oxychloride 0.25% or carbendazim 0.1% affords protection to the emerging seedlings.

General characters of genus *Rhizoctonia* - Refer page - 140

vi) Basal wilt of pepper - *Sclerotium rolfsii*

Basal wilt of pepper

'Basal wilt' of pepper is a nursery disease and attacks mainly the seedlings. The symptoms appear as grayish lesions on the stems and leaves of seedlings. Fine, whitish mycelial growth is seen at the advancing edges of the lesions on the leaves and on the stems. The mycelial growth then girdles the stem resulting in drooping and wilting of leaves above the point of infection and as the disease advances the seedlings arising from the rooted cuttings dry up. Small, whitish to cream colored sclerotial bodies appear on the mature lesions, which turn into brownish, mustard seed-sized sclerotia.

The pathogen is soil-borne and the sclerotia can survive in the plant debris and soil for years in a viable state in the absence of hosts and cause primary infection when suitable hosts appear. The sclerotia are transported to other places through soil and running water.

Sclerotium rolfsii thrives in highly aerobic environments and so thrives best near the soil surface. High temperatures ranging from 27° - 35°C, humid conditions and acidic soils are favorable for the fungal growth and disease development.

Control measures

- Plant debris should be removed from the soil and destroyed.
- Nurseries should be established in fields not contaminated by the pathogen.

- Alternate and collateral hosts of the pathogen, which may harbor the pathogen should be removed from the nursery area as well as from areas round about the nursery area.
- Drenching the cuttings with carbendazim 0.1% or Bordeaux mixture 1.0% or copper oxychloride 0.25% at the time of planting affords protection to the seedlings.

General characters of genus *Sclerotium* - Refer page - 141

vii) Stunt disease of pepper - A strain of cucumber mosaic virus

The disease also known as 'Little leaf', 'Wrinkled leaf' and 'Sickle leaf disease' is found in almost all the pepper growing countries. The leaves of infected plants become small and narrow with varying degrees of deformation and appear leathery, puckered and crinkled. Chlorotic spots and streaks may also appear on the leaves. The affected vines show shortening of internodes. The yield is also badly affected.

Stunt disease of pepper

The disease is mainly transmitted through infected stem cuttings used for propagation. The virus is transmitted by aphids and mealy bugs.

Control measures

- Disease-free, healthy planting materials should be used
- Severely affected vines may be removed and destroyed to prevent spread of the disease
- The insect vectors should be controlled by spraying dimethoate 30 EC 0.06% (2.0 ml/litre of water) or monocrotophos 36 SL 0.04% (1.0 ml/litre of water)

viii) Phyllody disease of pepper - Phytoplasma-like organism

'Phyllody' also known as **'Phyllomorphy'**, which is caused by a Phytoplasma-like microorganism attacks pepper crop sporadically. Affected vines exhibit general foliar yellowing and varying kinds of malformation of the spikes. The floral buds in such malformed spikes are transformed into narrow, leafy structures and exhibit phyllody symptoms. In advanced stages the leaves become small, twisted and chlorotic and the internodes of the vines are shortened.

Phyllody of pepper

The fruiting laterals are turned into leafy structures and present a witches broom appearance. Severely affected vines become unproductive within two to three years. The disease is mainly perpetuated through infected cuttings used as seed material. The causal organism is transmitted by leaf hoppers and plant hoppers.

Control measures

- As the disease is spread mainly through cuttings used as seed material, care should be taken to use disease-free, healthy planting material.
- Severely affected vines should be removed and destroyed
- All collateral and alternate hosts of the causal organism and the vectors in pepper fields and in the surrounding areas should be eliminated.
- The insect vectors should be controlled by spraying dimethoate 30 EC 0.06% or monocrotophos 36 SL 0.04%. Two to three sprayings should be given at fortnightly or three weeks intervals.

3. Cardamom *(Elettaria cardamomum)*

"Cardamom' popularly known as **'Ilachi'** is native to South India and Guatemala. It is cultivated in countries like India, Nepal, Sri Lanka, Guatemala Mexico, Thailand, Indo China, Tanzania and Central America. Kerala, Karnataka and Tamil Nadu are the major cardamom growing states in India.

Cardamom known as the **'Queen of spices'** belongs to the Family - Zingiberaceae. Cardamom plant is a perennial, flowering herb with thick fleshy rhizomes and leafy shoots with long green leaves. Cardamom is actually the dried ripe fruits (Capsules of cardamom plant). The ripe fruit is a green , three-sided, oval capsule containing 15 - 20, dark, reddish brown to brownish black, hard, angular, somewhat sticky seeds.

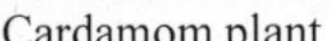

Cardamom plant

Cardamom pods

Cardamom is famous for its characteristic aroma and taste and is widely used as a flavoring and spicy ingredient in various kinds of curries and confectionary products. The seeds and oil from the seeds are used to make medicines capable of curing many ailments, such as digestion problems, intestinal spasms, diarrhea, constipation, liver and gall bladder complaints, loss of appetite etc. It is also used for common cold, cough, bronchitis, sore throat, urinary problems, epilepsy, headache, high blood pressure and even cancer.

Diseases of cardamom (*Elettaria cardamomum*)

i) 'Katte' disease/Cardamom mosaic disease / Marble disease of cardamom - Cardamom mosaic virus (CdMV)

'Katte' is the most destructive virus disease affecting cardamom plantations all over the world. The virus 'Katte mosaic virus' is spread through infected suckers and through the aphid vectors *Pentalonia nigronervosa* f.sp. *caladii* and the banana aphid *Pentalonia nigronervosa* f.sp. *typica.* The disease is systemic and the virus is present in all the parts of the plants except in the mature seeds. Both nymphs as well as winged and non-winged adult aphids can transmit the virus in a non-persistent manner. Some of the common weed plants found in cardamom gardens, such as *Caladium* sp., *Colocasia* sp., *Alpinia* sp. etc. serve as alternate hosts for the virus. The first visible symptoms appear on the youngest leaf of the affected tiller as spindle-shaped, slender, chlorotic, interrupted flecks. Later these flecks develop into pale green discontinuous stripes, which run parallel to the veins from the middle to the leaf margin. Subsequently all the newly emerging leaves show characteristic mosaic symptoms. Newly formed leaves become small and the affected plants become very much stunted. When young plants are affected flower and capsule formation is almost completely stopped. In infected fully grown plants, the yield is drastically reduced. The disease spreads to all the tillers in a clump.

'Katte' disease of cardamom

Control measures

- Use of disease-free, healthy suckers for seed purposes, removal of alternate and collateral hosts, such as *Colacasia* and *Caladium*, removal of disease affected clumps etc. are helpful in reducing the disease incidence.
- Spraying the crop with dimethoate 0.06% (2.0 ml / litre of water) or monocrotophos 0.04% (1.0 ml / litre of water) or phosphamidon 0.05% (0.5

ml / litre of water) or methyl demeton 0.05% (2.0 ml / litre of water) or neem based products at 0.1 per cent has been found to be effective in controlling the aphid vector.

ii) Damping off of seedlings / Seedling rot of cardamom - *Pythium vexans* and *Rhizoctonia solani* (Perfect stage - *Thanatephorus cucumeris*)

'Damping off of seedlings' or **'Seedling rot'** is a soil-borne disease that affects the seeds and new, emerging seedlings. The disease causes rotting of the stem and root tissues at and below the soil surface. The affected seedlings become water soaked and mushy, topple over at the base and die in masses. The disease appears during the monsoon season when there is excessive soil moisture.

Control measures

- Thick sowing should be avoided
- Pre treatment of seeds with *Trichoderma harzianum* or *T. viride* or *Pseudomonas fluorescens* before sowing reduces the chances of early infection in the nursery. Soil application of *Trichoderma* spp. in the nursery bed at 100 g / m ^{2} reduces further disease spread.
- Affected seedlings should be removed and destroyed by burning so as to avoid build up of inoculum in the soil.
- When initial symptoms of the disease is noticed drench the nursery beds with Bordeaux mixture 1.0% or copper oxychloride 0.25% at the rate of 3 - 5 litres per square meter. Subsequently two or three drenching are to be given at 15 days interval

General characters of genus Pythium - Refer page - 133

iii) Primary nursery leaf spot of cardamom - *Phyllosticta elettariae*

Primary nursery leaf spot

Disease symptoms appear as small, round or oval spots, which are dull white in color. These spots later become necrotic and the central portion breaks down and falls off leaving a hole thus exhibiting shot hole symptom. The spots are surrounded by water soaked area. When the disease incidence is severe numerous spots may develop on the leaves. Severely affected seedlings may collapse and die. High relative humidity and persistent dew favor occurrence and spread of the disease. The disease is mainly air-borne and the pycnidiospores released from the pycnidia are carried by wind and cause secondary infection.

Control measures

- Severely affected seedlings should be removed and destroyed .
- Spraying with copper oxychloride 0.25% or carbendazim
- 0.1% or mancozeb 0.2% solution is effective in controlling the disease.

General characters of genus Phyllosticta - Refer page - 135

iv) Secondary nursery leaf spot / Cercospora leaf spot of cardamom - Cercospora zingiberi

The disease occurs both in the nursery as well as in the plantations. The symptoms appear as yellowish to reddish-brown colored, rectangular patches running almost parallel to the veins. In advanced stages the color of the lesions turn into muddy red.

Secondary nursery leaf spot

The disease is air-borne and the conidia produced from conidiophores arising from the diseased patches are carried by air currents and cause fresh infection. The disease usually appears with the receipt of summer showers. Optimum temperature range for the growth of the pathogen is 24°C.

Control measures

Spraying the crop with mancozeb 0.2% or copper oxychloride 0.25% or Bordeaux mixture 1,0% affords good control of the disease. Fungicidal application should be repeated two or three times at fortnightly intervals.

General characters of genus *Cercospora* - Refer page

v) Capsule rot / 'Azhugal' disease of cardamom - *Phytophthora nicotianae* var. *nicotianae* and *P.meadii*

'Capsule rot' of cardamom is a serious problem in cardamom plantations and poses a major constraint in the successful cultivation of cardamom crop. Symptoms appear on the leaves as water-soaked lesions and is followed by rotting and shredding of leaves along the veins. Infected capsules become dull greenish brown and decay. Decayed capsules emit a foul odor and are shed subsequently. Infection spreads to the panicles and tillers and they decay.

Capsule rot of cardamom

The disease appears during the rainy season. Continuous rainfall and high relative humidity are favorable for the disease occurrence and spread.

Control measures

- Spraying with Bordeaux mixture 1.0% or copper oxychloride 0.25% or mancozeb 0.2% before the onset of south west monsoon rains followed by two or three sprayings afford fairly good control of the disease.
- Spraying with fosetyl Al. 0.1% gives effective control of the disease. A combination of fosetyl Al and mancozeb 0.1% affords better control of the disease.

General characters of genus *Phytophthora* - Refer page - 142

vi) Leaf blight of cardamom - *Phytophthora meadii*

Besides causing 'Azhugal' disease of cardamom, *Phytophthora meadii* also causes leaf blight disease of cardamom. The infection starts on the young, middle-aged leaves in the form of elongate or ovoid, large, brown-colored patches, which soon become necrotic and dry. The necrotic dry patches are seen mostly on leaf margins and in severe cases the entire leaf area on one side of the mid rib is affected.

Leaf blight of cardamom

The disease, which appears during the North East monsoon periods, becomes severe during the late monsoon periods from October to November. Intermittent rains and prevalence of misty conditions predispose the plants to infection and spread. The sporangia produced by the pathogen require water droplets on the leaf surface for germination and release of zoospores to cause fresh infection. Cardamom plants being perennial, the pathogen can continue their life in the plants and produce sporangia when conditions turn favorable for reproduction.

Control measures

Spraying with copper oxychloride 0.25% or mancozeb 0.2% or carbendazim 0.1% affords good control of the disease. Spraying should be taken up as soon as initial symptoms of the disease appear followed by two or three more spraying at 30 days interval.

General characters of genus *Phytophthora* - Refer page - 142

vii) 'Chenthal' disease / Leaf blight and Capsule brown spot / Anthracnose of cardamom - *Colletotrichum gloeosporioides;* (Perfect stage - *Glomerella cingulata*)

'Chenthal' disease is one of the major fungal diseases and is of great concern in cardamom plantations. The disease appears during the mid south west monsoon period and becomes severe during the late monsoon period.

'Chenthal' disease of cardamom

The disease symptoms appear initially on the leaves as small, water-soaked, rectangular lesions, which later elongate to form parallel streaks. The lesions turn yellowish brown to orange red in color with a necrotic centre. As the disease advances more number of lesions are formed on the young as well as older leaves. Severely affected leaves dry up and the plant presents a burnt appearance. Severely affected plants produce poorly developed flowers and capsule setting is adversely affected.

The symptoms of capsule brown spot appear as small, round, reddish-brown lesions on the pericarp of the capsules. On maturity of the capsules, these spots become soft, reddish-brown, sunken areas. The red color of the spots is retained even after curing of the capsules. Intermittent rains and prolonged misty weather are conducive for the occurrence and spread of the disease.

Control measures

- Collection and destruction of blight affected leaves and plant debris before the onset of monsoon rains and providing adequate shade for the plants help to mitigate the disease incidence .
- Spraying the crop with Bordeaux mixture 1.0% or copper oxychloride 0.25% or mancozeb 0.2% or carbendazim 0.1% or a combination of carbendazim 0.1% and mancozeb 0.2% when initial symptoms of the disease appear, followed by two or three sprayings at monthly interval has been found to afford adequate control of the disease.

General characters of genus *Colletotrichum* - Refer page - 136

viii) Root tip rot of cardamom - *Fusarium oxysporum*

'Root tip rot' of cardamom is an important soil-borne disease that affects the root system of cardamom plants.

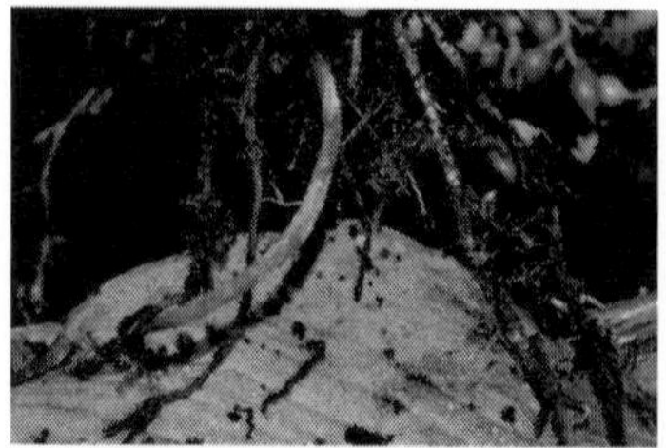
Root tip rot of cardamom

The symptoms are manifested as rotting of root tips followed by die-back of roots. The lower leaves of affected tillers turn yellowish and gradually dry. Such affected tillers become weak at this portion, break down partially and hang down.

The disease occurs during the post monsoon period. Relatively high soil moisture and soil temperature predispose the plants to disease occurrence and spread. The disease spreads through soil and cultural implements.

Control measures

- Spraying and drenching the plant basins with carbendazim 0.1% or hexaconazole 0.2% are recommended and the treatments have to be repeated at 10 -15 days interval.

General characters of genus *Fusarium* - Refer page - 138

ix) Cardamom necrosis / Nilgiri necrosis - Rod-shaped virus

'Nilgiri necrosis' virus disease occurs only sporadically in some localized areas. The symptoms appear as alternate light green and whitish to yellowish continuous or broken streaks on the leaves in the form of mosaic. Later these stripes become necrotic and turn reddish-brown in color. This is followed by tearing of the leaf lamina along the stripes. The leaves are wrinkled and the leaf margins become wavy. The necrotic areas soon dry off. The tillers become stunted, narrow and they produce small leaves, which are brittle. Panicles become shorter and produce only a few capsules, which show cracks.

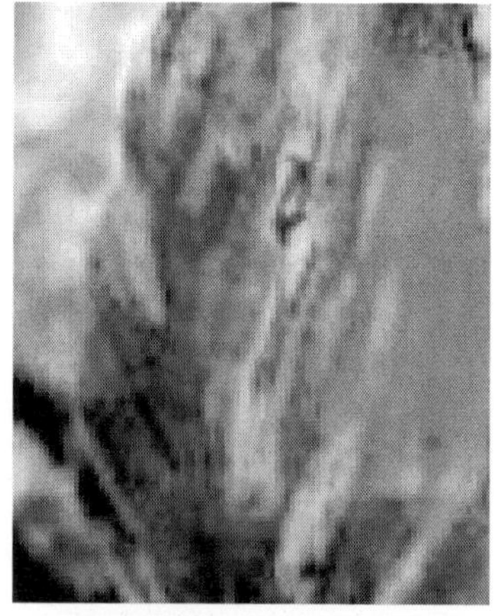
Nilgiri Necrosis of Cardamom

The disease caused by a rod-shaped virus is systemic and the virus is found in all the parts of the plants. The disease is transmitted only through infected rhizomes.

Control measures

- Severely affected plants should be removed and destroyed.
- Only disease-free, healthy seedlings raised in disease-free areas should be used for planting.

x) Clump rot / Rhizome rot of cardamom - *Pythium vexans, Rhizoctonia solani and Fusarium* spp.

Clump rot of cardamom

Early symptoms of the disease appear on leaves as pale yellow patches along the leaf margins and withering of leaves. Rotting or decay starts at the collar region of tillers and spreads to the rhizomes and roots. Rotting at the collar region results in toppling and death of tillers. Affected tillers can be pulled out quite easily with a slight pull and discoloration and rotting of the basal portion of the clump can be seen. Rotten rhizomes become soft, dark brown colored and emit a foul odor.

The disease is mainly soil-borne. The pathogens can survive in the plant debris and soil indefinitely.

Low temperatures below 24°C for a few days is ideal for infection and development of the disease. High humidity, high soil moisture, cloudiness and low temperatures predispose the plants to infection and spread of the disease.

Control measures

- Removal and destruction of plant debris from the soil, destruction of alternate and collateral hosts of the pathogens, providing adequate drainage facilities, removal and destruction of severely affected clumps, are measures that can be followed to reduce infection and spread of the disease.
- Treating the rhizome seed pieces with *Trichoderma harzianum* at 5.0 g / Kg of seed rhizomes, soil application of *Trichoderma harzianum* in the nursery at 100 g / m^2 has been found to be effective in reducing the inoculum in the soil.
- Drenching the soil with copper oxychloride 0.25% or Bordeaux mixture 1.0% solution at the time of planting followed by two to three more drenching at 30 days interval affords good control of the disease.

xi) Cardamom vein clearing or *'Kokke Kandu'* disease of cardamom - Cardamom vein clearing virus

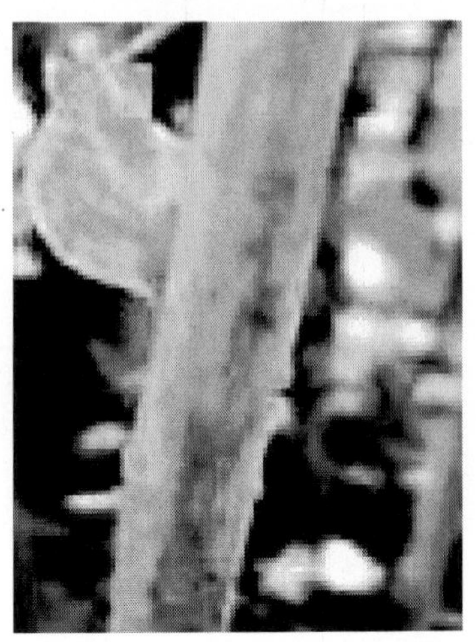
Cardamom vein clearing

The symptoms appear as continuous or discontinuous interveinal clearing, stunting and rosetting of leaves, loosening of leaf sheath, shredding of leaves and clear mottling on stem. In advanced stages of infection, the tips of tillers form hook-like structures as the younger most leaf roll gets hooked up in the leaf sheaths. This is the main characteristic symptom of the disease and is locally called as **'kokke kandu'**. Clear light green patches with three shallow grooves are found on the immature capsules. Cracking of capsules and partial sterility of seeds are the other symptoms of the disease.

The disease is mainly transmitted through infected rhizomes used as seed material. Secondary spread of the disease is through the vector cardamom aphid *(Pentalonia caladii)* in a non-persistent manner. The virus is also transmitted by the banana aphid (*Pentalonia nigronervosa*).

Control measures

- Use of disease-free, healthy rhizomes as seed material and use of healthy seedlings raised from disease free planting materials, removal and destruction of severely affected plants are methods that can be followed to prevent occurrence of the disease.
- The aphid vectors can be controlled by application of systemic insecticides, such as dimethoate 30 EC 0.06% (2.0 ml / litre of water) or methyl demeton 25 EC 0.05% (2.0 ml / litre of water) or phosphomidon 85 WSC 0.05% (0.5 ml / litre of water).

4. Clove *(Syzygium aromaticum)*

"Cloves' are the aromatic flower buds of the clove tree belonging to Family - Myrtaceae. Clove tree is a medium-sized, evergreen tree with large leaves and sanguine flowers grouped in terminal clusters. The small reddish-brown flower bud resembles a small nail head. Cloves are actually the dried, unopened flower buds.

Clove is native to the Maluku islands in Indonesia. Besides Indonesia, clove spice is grown in Zanzibar, Madagascar and Tanzania. It is also grown in India,

Sri Lanka and Pakistan. Karnataka, Kerala and Tamil Nadu are the main clove growing states in India. Nearly 80% of the clove of the world is produced by Indonesia.

Clove tree

Clove is one of the most valuable spices that has been used as a food additive and food preservative. A pungent, flavoring and culinary spice, it is found in a number of savory dishes, desserts, soups, stews and drinks. Ground or whole cloves flavor meat, sauces and rice dishes. Cloves are also used in spice cookies and cakes. Besides food items, cloves are also used in cosmetics, tooth pastes and medicines. Cloves are used in a type of cigarette called **'Kretek'** in Indonesia.

Cloves have been used tonically in traditional Chinese medicines and Ayurvedic medicines to strengthen the immune system, reduce inflammation and also to aid digestion. Cloves contain up to 18% of clove oil and about 89% of clove oil is eugenol. Eugenol present in clove oil is a powerful germicide. Besides its use as an antimicrobial agent, clove oil is also used to kill parasites and to repel insects. Clove oil is considered to be a best remedy for tooth ache and dental pain. Clove oil can protect against cancer by killing the cancer cells. It gives protection against stomach ulcers and boosts digestion. It also promotes skin health. Eugenol present in clove oil serves as a good food preservative.

Diseases of clove (*Syzygium aromaticum*)

i) Leaf rot / Leaf blight of clove - *Cylindrocladium quinquiseptatum*

The foliar disease occurs both in the nurseries and in the main fields and all growth stages of the plants are subjected to attack. Initial symptoms appear as dark spots at the leaf margin, which spread to all the leaf area without any

Branches with buds

Dried clove

definite pattern. Subsequently the entire affected leaf may rot or the tip portion of the leaf may rot resulting in defoliation.

The fungus overwinters as microsclerotia, which can survive for long periods of time, even for many years in soil or on infected plant tissues and cause infection. Secondary infection is caused by conidia spread through water or air. High relative humidity and moderate temperature favor disease occurrence and spread.

Control measures

- Spraying the crop with mancozeb 0.2% or carbendazim 0.1% or a combination of mancozeb + carbendazim has been found to be effective in providing adequate control.

General characters of genus *Cylindrocladium* - Refer page - 145

ii) Leaf spot, Twig blight and Flower shedding of clove - *Gloeosporium (Colletotrichum) gloeosporioides*; (Perfect stage - *Glomerella cingulata*)

Initial symptoms appear as small, circular to oval, brown colored, scattered specks on the leaf lamina. The specks gradually enlarge to form distinct spots with gray centre and dark margin. These spots coalesce and develop into necrotic patches. Infection spreads to the petioles leading to defoliation. On young twigs scattered small, brown spots appear, which coalesce and develop into large necrotic spots leading to die-back and death of twigs. Small black spots appear on flower buds and flowers, the petioles dry and the flower buds and flowers are shed in large numbers. Young flower buds are more vulnerable to attack by the pathogen.

The causal organism has a wide range of alternate and collateral hosts. The disease is mostly air-borne through conidia.

Heavy and continuous rainfall favor disease occurrence and spread. Moist weather conditions during the bloom and moderate temperature are conducive for development of the disease.

Control measures

- Collateral hosts, such as *Clerodendron* growing in the clove gardens should be removed and destroyed.
- Spraying the crop with Bordeaux mixture 1.0% or copper oxychloride 0.25% or mancozeb 0.2% when initial symptoms appear on the foliage affords control of the disease. The spraying has to be repeated at fortnightly intervals.

General characters of genus *Gloeosporium* - Refer page - 136

iii) Die-back / Eucalyptus canker of clove - *Cryphonectria cubensis*

Canker of Clove

'Die-back' of clove otherwise known as **'Eucalyptus canker'** is another serious disease of clove. The pathogen enters clove tree only through wounds. Once the fungus enters the host tree, it travels downwards until it reaches the branch junction. This is followed by drying and death of all the branches above the junction from the tip downwards resulting in die-back.

The fungus is a wound parasite and can enter the host only through wounds caused during cultural operations, harvesting or by natural means.

Control measures

- Causing damage to the trees by machinery and tools should be avoided.
- Pruning wounds should be treated with suitable fungicides, such as Bordeaux paste.
- Severely affected branches should be pruned and destroyed. The resultant wounds should be treated with fungicides.

General characters of genus *Cryphonectria* - Refer page - 146

iv) 'Sumatra disease' / Die back of clove - *Ralstonia syzygii* sub. Sp. *syzygii*

'Sumatra' disease of clove

'Sumatra' disease is considered to be the most important threat to clove production in Indonesia, where cloves are used in the manufacture of a cigarette called **'Kretek'**. The disease usually attacks the older trees above 10 years old, but younger trees are also subjected to attack. Symptoms appear as a progressive wilt and die back starting from the crown of the tree, which presents a 'stag's head' appearance. The affected trees die within a period of 6 months to 3 years. Leaves become chlorotic and fall off prematurely or may wilt and remain attached to the branches presenting a scorched appearance to the tree. Affected twigs turn brown and die back. The root system shows signs of degeneration and decay. Grayish brown streaks may be seen in the newly formed wood & a bacterial ooze may come out from the xylem vessels at the cut ends.

The bacteria causing the disease is spread by the insect vectors *Hindola striata* and *Hindola fulva*, sucking insect pests belonging to the Order- Hemiptera.

Control measures

- Use of antibiotics to control the disease has not been found to be successful.
- Use of insecticides to eradicate the insect vectors has been found to control the disease to some extent.

v) Sudden death disease of clove - *Valsa eugeniae*

'Sudden death' disease is one of the common diseases of clove. The disease primarily attacks mature clove trees. The symptoms of the disease appear as slight chlorosis of the leaves. This is followed by sudden and rapid leaf fall and wilting of the plants. The roots are also affected by the pathogen. The young roots decay and the wood under the bark of tree trunks develop an intense yellow color. Death of the affected plant occurs within a few days or a few weeks. The saplings and seedlings are not affected by the disease. The wound pathogen moves down from the point of infection and reaches the stem where it branches and forms cankers and the branches above the cankers die and results in die-back.

The pathogen produces large number of perithecia on the affected plant parts. The ascospores released from the asci in the perithecia are washed down to the soil and they infect the absorbing and fibrous roots and cause the disease by blocking the conducting vessels resulting in sudden death of the plants.

Rainy seasons favor the disease occurrence by promoting production of large numbers of ascospores and conidia.

Control measures

- Field sanitation, removal and destruction of alternate hosts, removal and destruction of affected plants are some of the cultural measures that can be followed to prevent occurrence of the disease. The branches below the infected area are cut and Bordeaux paste or copper oxychloride paste should be applied to the cut ends.
- There are no chemical control measures to control the disease.

Genera characters of genus *Valsa* - Refer page - 147

vi) Seedling wilt of clove - *Fusarium* sp., *Rhizoctonia* sp. and *Colletotrichum* sp.

Several species of fungi are involved in causing seedling wilt. The disease is found mainly in the nurseries. Leaves of affected seedlings loose their natural lustre, tend to droop, collapse from the collar region and die in large numbers in patches. The roots and collar region of the seedlings show varying degrees of discoloration and decay.

The pathogens are soil-borne and can continue to live in the soil for prolonged periods. The spores produced y the fungi in the infected seedlings also cause secondary infection. The spores produced by the pathogens are carried through soil, air and irrigation water to other fields and cause fresh infection.

High moisture content of the soil, high relative humidity and moderate temperatures favor the disease occurrence and spread.

Control measures

- Field sanitation, removal and destruction of diseased seedlings, providing proper drainage facilities are measures that can be taken to avoid the incidence of the disease.
- Drenching the soil with Bordeaux mixture 1.0% or copper oxychloride 0.25% after removing the affected seedlings helps to arrest further spread of the disease.
- Spraying the seedlings with Bordeaux mixture 1.0% or copper oxychloride 0.25% or carbendazim 0.1% and drenching the soil with any one of the fungicides mentioned above affords protection to the seedlings.

vii) Little leaf disease of clove - Little leaf Phytoplasma

The disease attacks both grown up plants as well as seedlings in the nurseries. Infected plants produce tiny, pale green leaves on very short petioles. Stem internodes are short and the plants are stunted and bushy due to stimulation of axillary buds. The primary and secondary branches are arranged closer to each other presenting a bushy appearance. Flowers formed are leaf-like and sterile. Root proliferation is also seen.

The causal organism is transmitted by the leaf hopper *(Hishimonus phycitis)*. Overlapping of clove crop cycling, presence of several weed hosts and crop hosts, such as egg plant encourage the survival of the vector all through the year and carry the phytoplasma and spread the disease. Grafting can also spread the disease.

Control measures

- Removal and destruction of affected plants, eradication of weeds and other hosts of the vector, application of suitable insecticides to destroy the vectors are measures that can be taken to prevent spread of the disease.
- The disease can be kept under check by spraying tetracycline.

5. Turmeric *(Curcuma longa)*

'Turmeric' is a flowering plant of the ginger family - Zingiberaceae, the roots of which are used widely in cooking of various dishes, both vegetarian and non-vegetarian. The plant is a perennial, rhizomatous, herbaceous plant native to Indian sub continent and southeast Asia. It requires a temperature range between 20° - 30°C and well distributed annual rainfall. Turmeric is the spice that comes from the rhizomes of the turmeric plant.

Turmeric plants

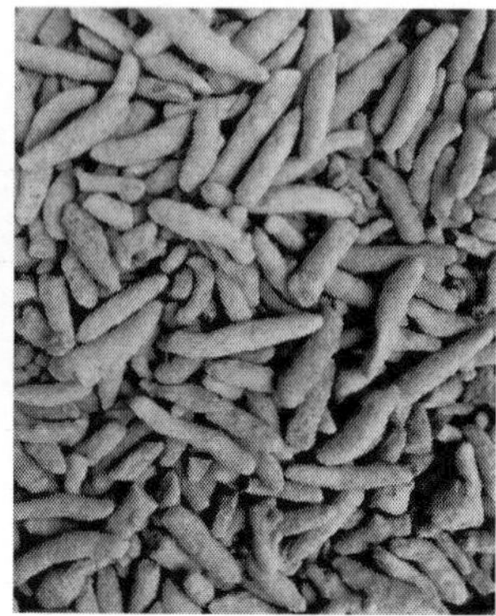

Dried turmeric rhizomes

Turmeric is used quite commonly in Asian food and is the main spice in curry. It has a warm, bitter taste and is frequently used to flavor or color various curry powders. It contains a yellow-colored chemical called 'curcumin', which is often used to color foods and cosmetics. The roots of turmeric is also widely used to make several medicines especially in Ayurvedic medicines.

As medicine, turmeric is commonly used for conditions involving pain and inflammation, such as osteoarthritis. It is also used for treatment of hay fever, depression, high cholesterol, liver complaints and itching. It is also being used for healing heart burns, inflammatory bowel diseases, stress, improving memory skills and many other conditions. It is widely used in Ayurvedic and Unani medicines. Curcumin and other potent chemicals present in turmeric decrease inflammation. Turmeric is also used in several religious rites and ceremonies.

Diseases of turmeric (*Curcuma longa*)

i) Rhizome rot of turmeric - *Pythium aphanidermatum*

'Rhizome rot' of turmeric is a serious disease of turmeric that causes heavy damage to the rhizomes. The disease is both soil-borne and seed-borne by means of seed rhizomes. The disease is mostly prevalent during the south west monsoon season.

The infection starts at the collar region of the pseudostem and progresses both upwards and downwards. Foliar symptoms appear as light yellowing of the tips of lower leaves and then spread to the leaf border Initially the middle portion of the leaves remain green, while the margins become yellow. Soon the yellowing

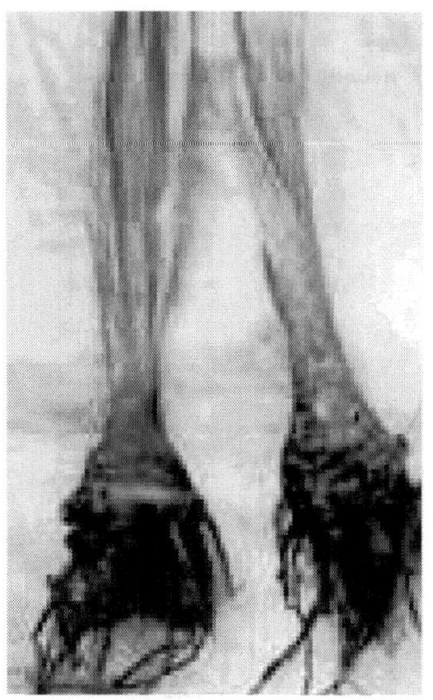
Rhizome rot of turmeric

spreads to all the leaves from the bottom to the top and is followed by drooping, withering and drying of the leaves, and the pseudostem. Infected rhizomes become rotten, soft and pulpy, and the color changes into different shades of brown. The root system is very much reduced. Usually the disease appears in patches in the field.

The disease is primarily soil-borne. The pathogen survives in the soil as chlamydospores and oospores that get access to the soil from infected rhizomes and cause initial infection. Infection also occurs through the pathogen present in rhizomes used as seed material. At the onset of the south west monsoon, with build up of soil moisture the pathogen multiplies rapidly. Younger sprouts are more vulnerable to attack by the pathogen. Nematode infection aggravates the disease.

Temperature above 30°C, high soil moisture, water logging and poor drainage are favorable for disease occurrence and spread. Presence of abundant decomposing organic matter in the soil is another factor that contributes to the severity of the disease.

Control measures

- Selection of good quality, disease-free planting material, provision of adequate drainage facilities to avoid water stagnation, removal of disease affected plants, crop rotation, deep ploughing during the summer are measures that can be followed to reduce disease incidence.
- Steeping the seed rhizomes in copper oxychloride 0.25% or mancozeb 0.2% solution for 30 minutes and shade drying before sowing helps to minimize seed inoculum.
- Drenching the soil around the infected plants with mancozeb 0.3% or Ridomil MZ 0.2% or copper oxychloride 0.25% reduces further spread of the disease.
- Spraying the crop with mancozeb 0.2% or carbendazim 0.1% affords adequate control of the disease.
- Soil application of *Pseudomonas fluorescens* at 2.5 Kg/ha has been recommended for the control of the disease.
- Cultivars, such as Suvarna, Suguna and Sudarsana have been found to be resistant to the disease.

General characters of genus *Pythium* - Refer page - 133

ii) Leaf blotch of turmeric - *Taphrina maculans*

Leaf blotch of turmeric

'Leaf blotch' of turmeric is an important disease of turmeric that causes severe damage to the crop leading to substantial yield loss. Symptoms appear as individual, small, oval or mostly rectangular shaped spots, which are 1.0 to 2.0 mm in width. Several such spots appear on both the surfaces of the leaves. However, the spots are considerably more on the upper surface. The spots are arranged in rows parallel to the veins. The spots coalesce to form bigger and irregularly-shaped lesions. Initially the spots which are pale yellow in color turn into dirty yellow to brown later on. The infected leaves present a reddish brown scorched appearance. However the affected plants are not killed.

Primary infection occurs on the bottom leaves from spores from dried fallen leaves. Secondary infection occurs from ascospores released in large numbers from the asci, which attack all the leaves.

The incidence of the disease is more during the monsoon periods. High soil moisture, temperature about 25°C and leaf wetness are conducive for the rapid spread of the disease. A few other species of *Curcuma* and *zingiber* are also attacked by the pathogen and the spores produced by these alternate hosts may also infect turmeric crop.

Disease Control

- Selection of seed material from disease-free areas, removal and destruction of infected and dried leaves and crop residues, crop rotation and field sanitation measures help to ward off the disease occurrence and spread.
- Steeping the seed material in Mancozeb 0.2% or Carbendazim 0.1% solution for 30 minutes and shade drying the same prior to sowing protects the crop from initial infection.
- Spraying the crop with Mancozeb 0.2% or Carbendazim 0.1% solution when initial symptoms are noticed, followed by 2 - 3 more sprayings at fortnightly interval has been found to afford adequate control of the disease.

General characters of genus *Taphrina* - Refer page - 149

iii) *Colletotrichum* leaf spot of turmeric - *Colletotrichum capsici*

'Leaf spot' of turmeric is one of the most important and destructive disease of turmeric crop and has become a major constraint in successful cultivation of

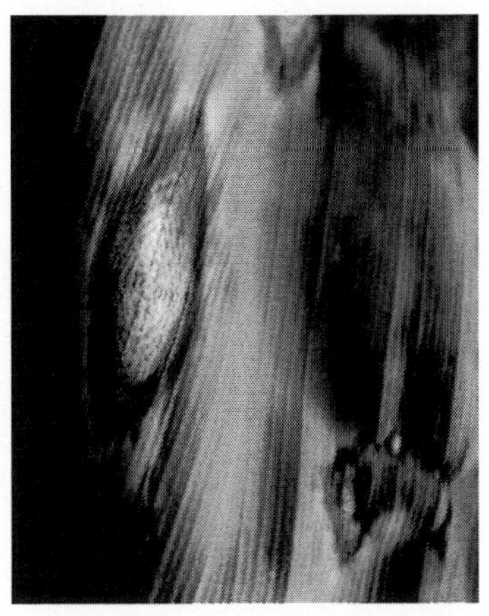

Colletotrichum leaf spot of turmeric

turmeric crop. The symptoms of the disease appear as small brown spots of different sizes, which develop into oblong brown spots with gray centre on the upper surface of young leaves. The spots are of size varying from 4.0 - 5.0 cm in length and 2.0 - 3.0 cm in width and are surrounded by a yellow halo. As the disease advances, several spots may coalesce, enlarge and cover major portion of the leaf blade. In advanced stages minute, pinhead like black dots representing the acervuli of the pathogen are formed in concentric rings in the central portion of the spots. Severely affected leaves wilt and dry. Spots may be formed on the leaf sheaths and the scales around the rhizomes.

The disease is more severe during the months of July to October. Relative humidity of 80% and temperature ranging from 21° - 23°C are favorable for the occurrence and spread of the disease.

Primary infection occurs from the fungus carried on the scales of rhizomes used as seed material. Secondary infection is caused by conidiospores released from the acervuli, which are spread by wind, water splash and by other physical means and biological agents.

Control measures

- Selection of seed material from disease-free areas, collection and destruction of infected and dried leaves, crop rotation with non-host crops wherever possible are measures that can be taken to prevent disease occurrence.
- The pathogen attacks chilli crop and causes leaf spot and fruit rot. So cultivation of chilli crop near about turmeric fields should be avoided. Including chilli crop in the crop rotation sequence should also be avoided.
- Steeping the seed material in mancozeb 0.2% or carbendazim 0.1% solution for 30 minutes and shade drying before sowing prevents primary infection.
- Spraying the crop with Mancozeb 0.2% or Carbendazim 0.1% or Copper oxychloride 0.25% or Topsin M 0.1% solution when initial symptoms appear followed by 2 - 3 more sprayings has been found to give adequate control of the disease.
- In endemic areas cultivation of resistant cultivars, such as 'Suguna' and 'Sudarshan' should be encouraged.

General characters of genus *Colletotrichum* - Refer page - 136

iv) Bacterial wilt of turmeric - *Ralstonia solanacearum*

'Bacterial wilt' disease causes rapid wilting and death of the entire plant without any yellowing or spotting on the leaves. All branches of the affected plant wilt and die simultaneously. The stem of a wilted plant when cut across appear water-soaked and dark colored and a grayish, slimy ooze comes out when the stem is pressed. In advanced stages of the disease the pith region decays completely and the stem becomes hollow.

The disease is mainly soil-borne. The bacteria causing the disease can live in the organic residues in the soil for indefinite periods. They enter the plants through wounds or injuries caused during cultural operations or by soil inhabiting nematodes and cause infection. The disease is also carried through infected rhizomes and infected seedlings.

High soil temperatures above 24°C and high soil moisture favor disease occurrence and spread.

Control measures

- Planting disease-free rhizomes or disease-free seedlings, judicious water management, eradication of alternate hosts, field sanitation, removal and destruction of severely affected plants are measures that can be followed to prevent infection and spread of the disease.
- Treating seed rhizomes in hot water at 47°C for 3 minutes and shade drying the same before planting eradicates the bacterial inoculum inside the rhizomes.
- Once the disease attacks the plants there is no curative measures.

v) Leaf blight of turmeric - *Rhizoctonia (Moniliopsis) solani*

This is a minor disease of turmeric and appears sporadically. The disease is characterized by the appearance of necrotic patches with papery white centre of varying sizes on the leaf lamina, which soon spread over the whole surface leaving a blighted appearance. White mycelial growth is seen on the stem region also and minute, black, pin-head like sclerotia are formed on the mycelial growth.

The pathogen is mainly soil-borne and the sclerotia can remain in a viable state for indefinite periods and cause initial infection. The pathogen is also carried in the rhizome seed material and perpetuate the disease. The disease mostly occurs during the post monsoon periods.

Control measures

- Field sanitation and selection of disease-free seed material are measures that can be followed to avoid the occurrence of the disease.

- Spraying the crop with Bordeaux mixture 1.0% or copper oxychloride 0.25% or carbendazim 0.1% or mancozeb 0.2% has been found to be effective in controlling the disease.

General characters of genus *Rhizoctonia* - Refer page - 140

vi) Brown rot of turmeric - *Pratylenchus* sp. and *Fusarium* sp.

This is another minor disease of turmeric and occurs sporadically. It is a complex disease caused by the nematode *Pratylenchus* sp. associated with *Fusarium* sp. Due to attack by the disease, the rhizome becomes deep gray to dark brown in color, less turgid, wrinkled and exhibit dry rot symptoms. On cutting open the affected rhizome, dark brown necrotic lesions starting from the margin into the internal tissues are seen.

The disease is soil-borne and both the nematode and the fungus can live in the soil as saprophytes and attack the crop when the crop is present. The fungus gets access to the crop through injuries caused by the nematode in the roots and in the rhizomes.

Once the disease appears it is rather impossible to control it Only preventive measures should be taken so as to avoid the occurrence of the disease.

6. Garlic *(Allium sativum)*

'Garlic' is native to central Asia and Iran. Now it is cultivated in all the countries of the world. China produces about 80% of the worlds production of garlic. Garlic has been used all over the world for thousands of years as a food additive and medicine.

Garlic is a monocotyledonous, herbaceous, annual, flowering, bulbous plant belonging to the Family - *Amaryllidaceae*

Garlic plants Garlic bulbs

Garlic is delicious and can be used in savory dishes, soups, sauces, dressings and in several other Indian and exotic food preparations, because of its pungent taste and delicious flavor. The compound 'Allicin' present in garlic is responsible for imparting the aroma in food preparations.

Besides its food values, garlic has many medicinal properties. Garlic purifies blood, relieves cold and flu, prevents heart diseases and cancer, reduces hypertension, lowers cholesterol levels, heals wounds, boosts digestion, regulates blood sugar, boosts immunity and controls Asthma.

Diseases of garlic (*Allium sativum*)

i) Damping off of garlic - *Pythium* spp., *Fusarium* spp., *Rhizoctonia solani* and *Sclerotium rolfsii*

'Damping off' of garlic may be caused by one or more than one organisms listed above. Pre-emergence damping off and post-emergence damping off may occur. In the case of pre-emergence damping off, the seeds as well as the germinating seedlings rot and die before they emerge out of the soil. In the case of post-emergence damping off, the collar region of the seedlings is attacked by the pathogens at the soil surface. The affected collar region rots and ultimately the seedlings collapse and die. The seedlings die suddenly in large numbers in patches.

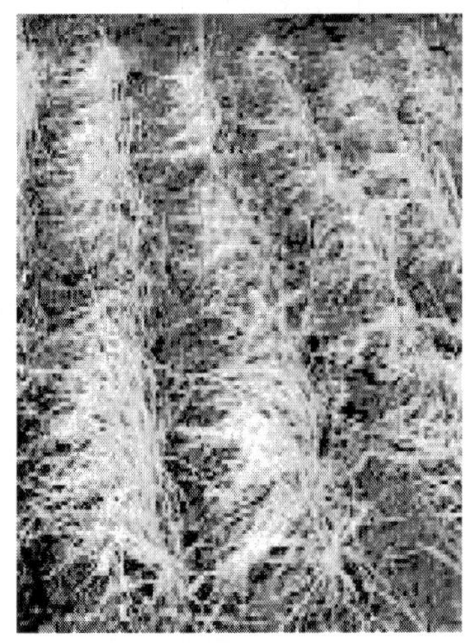

Damping off of garlic

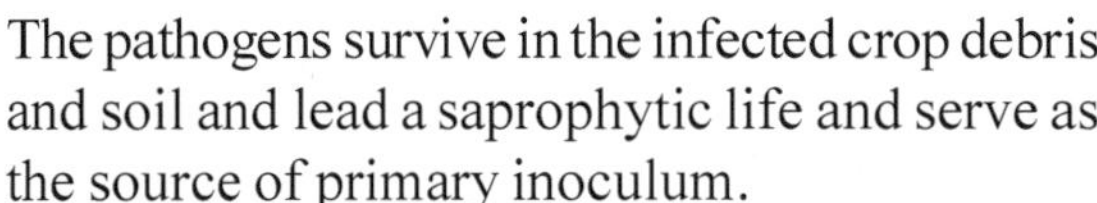

The pathogens survive in the infected crop debris and soil and lead a saprophytic life and serve as the source of primary inoculum.

The disease is more prevalent during the monsoon seasons

High soil moisture, high relative humidity and moderate temperature are conducive for the development of the disease.

Control measures

- Thick sowing should be avoided
- Soil application of *Trichoderma harzianum* or *T. viride* in the nursery bed at 100 g/m^2 reduces the disease incidence by eliminating the primary inoculum in the soil.
- Affected seedlings should be removed and destroyed by burning so as to avoid build up of inoculum in the soil.

- When initial symptoms of the disease is noticed drench the nursery beds with Bordeaux mixture 1.0% or copper oxychloride 0.25% at the rate of 3 - 5 litres per square meter. Subsequently two or three drenchings have to be given at 15 days interval.

ii) Downy mildew of garlic - *Peronospora destructor*

'Downy mildew' of garlic is one of the serious and destructive disease of garlic. The disease attacks the plants at all the growth stages of the crop. The symptoms often appear first on older leaves. Oval or cylindrical patches of varying sizes develop on infected leaves and the main stalk. These areas are pale greenish-yellow to brown in color. Masses of gray to violet, fuzzy, downy growth comprising of sporangiophores and sporangia develop on the infected leaves. Ultimately the leaves get girdled, collapse and die. The dead leaf tissues turn dark in color and the downy growth gets obscured. The bulb development is adversely affected and the bulb tissues and the neck region may become spongy.

Downy mildew of garlic

The pathogen can survive in left over bulbs in the field after harvest, volunteer plants, alternate hosts, such as onion, shallot, chive etc., as well as in discarded, diseased plant debris and produce spores and may cause primary infection. Perennating mycelium in the bulbs is the chief source of carry over of the disease Oospores adhering to the testa of seeds as well as the oospores that may be present in the crop residue in the soil may also cause primary infection. Secondary infection is through sporangia.

Cool, moist nights, high relative humidity above 95% and moderate warm day temperature favor disease occurrence and spread.

Control measures

- Use of disease-free seeds, setts and bulbs, removal and destruction of discarded, diseased plant debris, removal of alternate hosts, providing proper irrigation and drainage, crop rotation with non-host crops, removal of severely affected leaves and plants etc. are measures that can be followed to mitigate the disease incidence.
- Use of fungicides, such as copper oxychloride or dithiocarbamates can control the disease to some extent.

- Spraying a combination of metalaxyl 0.1% + mancozeb 0.2% has been found to be very effective in controlling the disease.

General characters of genus *Peronospora* - Refer page - 150

iii) White rot or *Sclerotium* rot of garlic - *Sclerotium cepivorum* (Perfect stage *Stromatinia cepivora*

The disease attacks the host during the active growth stages and continues to attack the bulbs in storage. Initial symptoms appear in older leaves, which turn yellow from the tips and proceed downwards resulting in die-back. Young plants are more severely affected resulting in complete wilting of all the leaves. The infection starts from the roots and then to the base of the bulb scales. In the later stages the bulbs are directly infected by the mycelium leading to rotting of the bulbs. The diseased plants can easily be pulled out from the soil. The rotten bulbs usually contain abundant white mycelium in the form of felt. On the mycelial weft, tiny, black, hard sclerotia are found in large numbers and appear as a black crust.

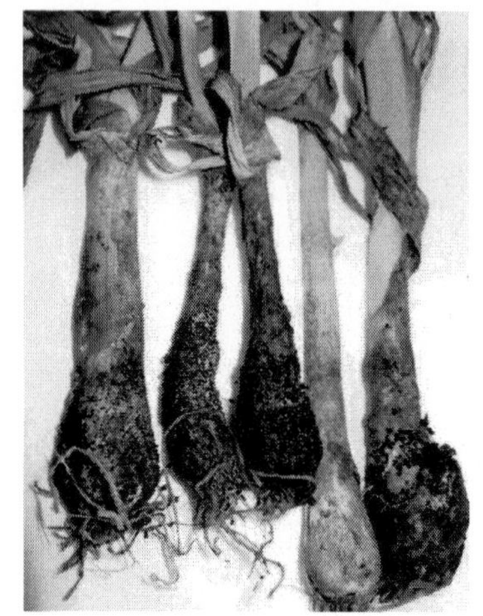

White rot of garlic

The pathogen can survive in the soil as sclerotia, which can remain in a viable state for 8 - 10 years. They are highly resistant to adverse soil conditions and can perpetuate the disease from one season to the ensuing seasons. The pathogen has many collateral and alternative hosts, such as onion, shallots, leeks, wild onion etc., which may also provide inoculum.

Cool weather conditions, low soil temperature between 15° - 18°C and moderate soil moisture favor disease occurrence and spread. The disease is almost absent at soil temperatures above 26°C and soil moisture content above 50%.

Control measures

- Removal and destruction diseased left over bulbs in the soil, diseased plant materials and debris from the fields, avoiding irrigation water to flow from disease affected field to adjacent fields, use of disease free and healthy planting materials, removal of alternate hosts from around garlic fields, adopting crop rotation practices, application of organic manure so as to encourage the growth of antagonistic micro organisms in the soil are measures that can be adopted to reduce the severity of the disease.

- Seed treatment with carbendazim or thiophanate methyl (Topsin M) at 2.0 g / Kg of seeds provides protection to the seedlings from soil-borne infection.

General characters of genus *Sclerotium* - Refer page - 141

iv) Bacterial soft rot of garlic - *Erwinia carotovora* pv. *Carotovora*

Bacterial soft rot of garlic

The pathogen attacks the bulbs in the field before harvest and after harvest in transit and storage. Initial symptoms appear as small, irregular, water-soaked lesions at the collar region. The infection spreads rapidly and progresses downward into the bulb. Soon after, all the tissues of the bulb are affected. The affected areas become soft and mushy. The outer scales rot completely and a slimy ooze is seen on the surface. When the infected bulb is pressed slightly, a turbid, slimy liquid oozes out, which gives off a repulsive odor.

The soil-inhabiting bacteria can survive in the soil and diseased plant debris for up to 20 years and lead a saprophytic life till they find a suitable host and then start a parasitic life. The bacteria are primarily wound parasites and can enter the host only through wounds caused during cultural operations or by insects, nematodes etc. Once the bacteria enter the host, they multiply enormously within a short period. They produce certain enzymes, which dissolve the cell walls as a result the tissues become a rotten, mushy mass. The bacteria are spread by direct contact, through agricultural implements, soil, water, insects and even by man.

High temperatures between 28° - 35°C and high soil moisture favor the occurrence and spread of the disease. Besides garlic the pathogen attacks several other crops, such as onion, carrot, turnip, radish, potato etc.

Control measures

- There are no specific control measures for controlling the disease. Because the bacteria are wound parasites, care should be taken to avoid causing injuries or bruises during cultural operations and while handling during harvest, transit and storage.
- Removing and destroying diseased plant debris, avoiding irrigation water flowing from diseased affected fields to other fields, sorting and removing

disease affected bulbs at the time of harvest, avoiding mixing up of diseased bulbs with sound bulbs, drying the bulbs sufficiently before storing, keeping the store houses clean, moisture free and well ventilated, adopting crop rotation practices are measures that can be followed to minimize disease occurrence.

v) Purple blotch of garlic - *Alternaria (Macrosporium) porri*

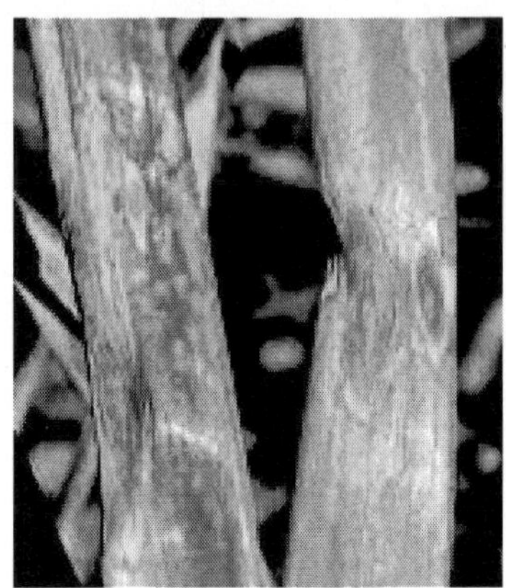
Purple blotch of garlic

The disease usually attacks the older leaves and flower stalks. Symptoms appear as small, sunken, whitish flecks with purple colored centre. The flecks enlarge into oval, brown to purple blotches with prominent yellow rings. Light and dark zonations develop on the blotches. The lesions may girdle the leaves and flower stalks and the leaves above the blotches wilt and die from the tip downwards. Younger leaves become more susceptible as the bulb matures. Bulbs may become infected through neck wounds and the bulb size is markedly reduced. The bulbs may rot in the field as well as during storage. Bulb rot symptoms appear as soft, water-soaked patches and turn dark reddish purple, and ultimately turn brown to black in color..The leaf tips are more prone to attack by the pathogen and the leaves wilt and die from the tip downwards.

The disease is soil-borne and the pathogen survives in infected bulbs in the soil, as well as in infected plant debris or on roots of weed hosts. The pathogen has a few other alternate hosts, such as onion, shallot, leeks etc. The spores (conidia) produced from such hosts may infect garlic crop.

Hot and humid climate with temperature ranging from 21° - 30°C and high relative humidity of 80 - 90% favor the development and spread of the disease. Infestation of onion thrips in garlic makes the plants weak, which also predisposes the plants to infection by the pathogen. Further the injuries caused by the thrips may pave the way for the entrance of the pathogen into the host and cause disease.

Control measures

- Use of disease-free planting materials for sowing, removal and destruction of infected plant debris from the fields, culling of volunteer plants that may carry the pathogen, eliminating weeds and other alternate hosts, harvesting the crop in dry weather, avoiding injuries to the bulb at the time of harvest and storage, adopting crop rotation practices, controlling thrips by use of suitable insecticides, curing the bulbs sufficiently before storage, storing the bulbs in clean, well ventilated ware houses etc. are measures that can be adopted to avoid disease incidence.

- Spraying mancozeb 0.2% or chlorothalonil 0.2% or iprodione 0.15% affords good control of the disease. Spraying should commence once the disease symptoms are noticed followed by 2 - 3 more sprayings at fortnightly intervals. Spraying the systemic fungicide triazole 0.1% has been found to give very good control of the disease.

General characters of genus *Alternaria* - Refer page - 153

vi) Black mould / Aspergillosis of garlic - *Aspergillus niger*

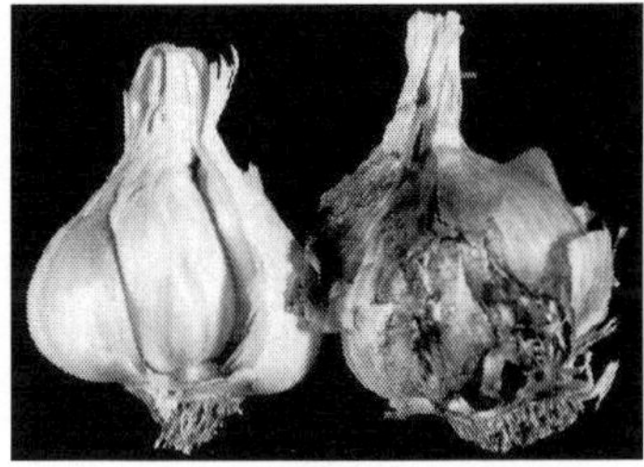
Black mould of garlic

The disease attacks garlic bulbs in the field as well as in storage. Infection occurs through the neck tissues as the foliage dies down at maturity. Infected bulbs become black around the neck region and the affected scales shrivel. Masses of powdery, black fungal spores (conidia) are formed as streaks along the veins and between outer dry scales. Infection may then advances from the neck into the central fleshy portions. Infection may also occur through wounds caused during cultural operations or while harvesting or during transit and storage.

Spores of the fungus is found everywhere floating in the air and may infect the bulbs. The fungus also survives in the soil and infected plant debris.

High temperatures above 30°C and humid conditions favor disease occurrence and spread. Free moisture on the surface of garlic bulbs predispose the bulbs to infection.

Control measures

- Avoiding mechanical damage to the bulbs at the time of harvest, transit and storage, proper drying of the bulbs before storing, storing the bulbs in clean, well ventilated ware houses and adopting such hygienic measures help to ward off infection by the fungus.

General characters of genus *Aspergillus* - Refer page - 154

vii) Onion yellow dwarf of garlic - Onion yellow dwarf virus

'Yellow dwarf' disease is caused by 'Onion yellow dwarf' virus., a Potyvirus that has a narrow host range comprising of Onion, garlic, shallot, leek and a few ornamental plants.

The symptoms appear as yellow stripes on mature leaves. Leaves show mosaic pattern discoloration accompanied by wrinkling, softening, curling and wilting of the leaves. The affected plants are stunted in growth and finally the entire plant turns yellow and wilt. Bulbs usually remain solid but are reduced in size.

The virus is carried by infected seed bulbs, volunteer plants and other alternate hosts. Many aphid species transmit the virus from diseased to healthy plants in a non-persistent manner. The virus is also sap transmissible.

Onion yellow dwarf of garlic

The virus is not seed-borne. So, use of true seeds promotes a healthy crop. Use of virus-free bulbs and setts for planting, raising the crop in fields where the disease was not prevalent, rouging out infected plants, adopting crop rotation practices, eliminating alternate hosts etc. help in preventing the occurrence of the disease. Controlling the aphid vectors by the application of suitable insecticides helps in preventing the spread of the disease to some extent.

viii) *Stemphylium* leaf blight of garlic - *Stemphylium vesicarium*

The disease occurs on leaves of transplanted seedlings at the 3 - 4 leaf stage during the summer months. The symptoms appear as scattered, small, yellowish to orange colored flecks or streaks in the middle region of the leaves , which soon develop into elongated, spindle-shaped spots surrounded by pinkish margin. The spots turn gray at the centre and later become brown to dark brown with the appearance of conidiophores and conidia of the pathogen. Finally the entire foliage is blighted. Similar symptoms appear on the inflorescence stalks also.

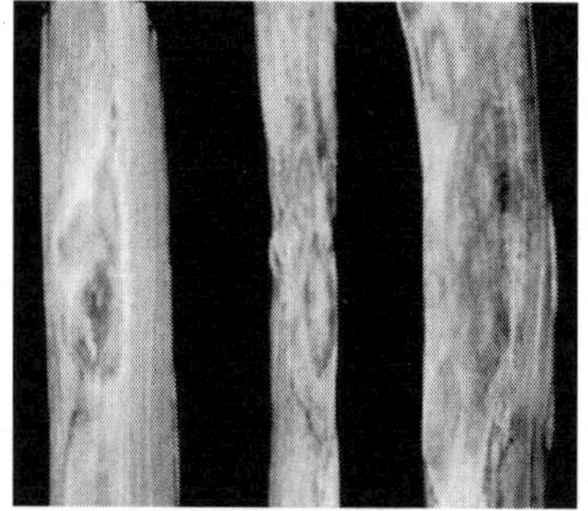

Stemphylium leaf blight of garlic

The fungus is soil-borne and is capable of surviving in the debris and soil. The spores produced from the pathogen are splashed by rain drops and infect the lower leaves. Secondary infection takes place through conidia carried by air currents. The fungus has several alternate and collateral hosts, such as onion, soybean, lucerne, asparagus etc. and the spores produced from such hosts also perpetuate the disease.

Moderate temperatures between 18° - 25°C, high humidity and long period of leaf wetness are favorable for disease occurrence and spread.

Control measures

- Removal and destruction of debris from the fields, eradication of alternate and collateral hosts near about garlic fields, crop rotation, provision of proper drainage facilities are measures that can be followed to prevent disease occurrence and spread.
- Foliar spraying with carbendazim 0.1% or chlorothalonil 0.15% or propiconazole 0.1% has been reported to give adequate control of the disease.

General characters of genus Stemphylium - Refer page - 156

ix) *Colletotrichum* blight/Anthracnose/Twister disease of garlic - *Colletotrichum gloeosporioides*

The disease attacks garlic crop at all growth stages. Initial symptoms appear on the leaves as small, water-soaked, pale yellow spots. The spots enlarge and spread length wise and cover the entire leaf blade. Curling and twisting of leaves and abnormal elongation of the necks are also seen.. The affected leaves shrivel, droop down and die prematurely. The affected plants produce slender bulbs. Some of the diseased plants rot before harvest.

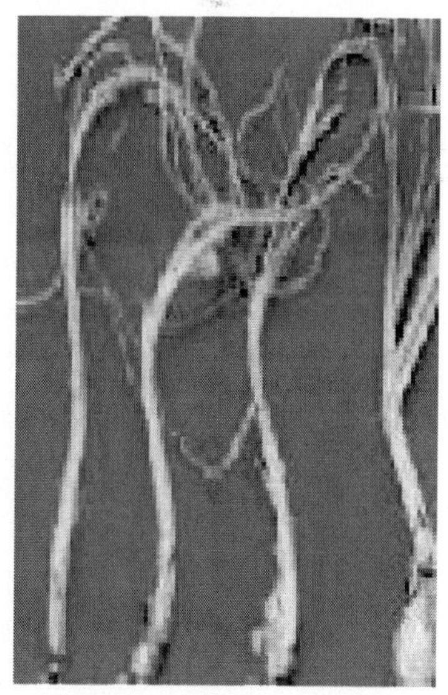

Twister disease of garlic

The pathogen is soil-borne and air-borne. The fungus remains viable for many years as sclerotia and for shorter periods in infected plant debris and causes initial infection. Secondary infection is caused by air-borne conidia produced from the acervuli in the host plants. The pathogen has many alternate and collateral hosts, such as onion, tomato, pepper, leek etc. and the conidia produced in these hosts can spread the disease to garlic.

The disease occurs mostly during the rainy seasons and is more severe when warm temperatures (25° - 30°C) and humid conditions prevail. Moist soils that are rich in organic matter encourages the disease development and spread.

Control measures

- Removal and destruction of debris from the soil, eliminating alternate hosts from garlic fields and nearby areas, proper water management, removal and destruction of severely affected plants, adopting crop rotation practices etc. are methods that can be followed to prevent occurrence and spread of the disease.
- Spraying the crop with Bordeaux mixture 1.0% or copper oxychloride 0.25% or mancozeb 0.2% or carbendazim 0.1% , when initial symptoms of the disease appear followed by 2 - 3 more sprayings at fortnightly intervals has been recommended.

General characters of genus *Colletotrichum* - Refer page - 136

x) *Fusarium* basal rot / Basal rot of garlic - *Fusarium oxysporum* f.sp. *cepae* and *Fusarium culmorum*

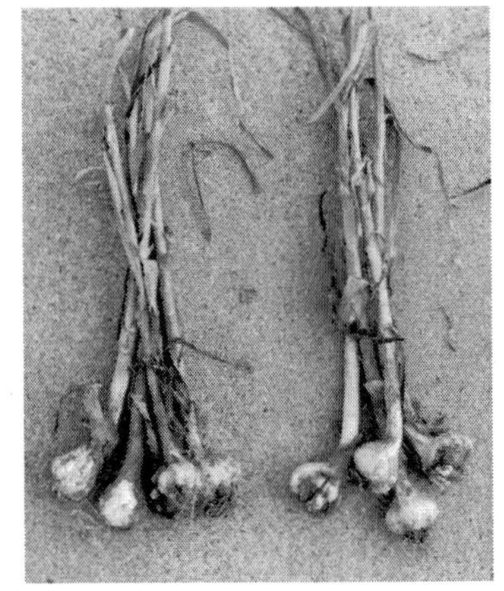

Basal rot of garlic

The disease appears mostly during the summer months and affects garlic crop at any growth stage. Symptoms appear as curling, yellowing and die back of leaves starting from the tips and develop downwards resulting in wilting of the leaves. Reddish brown discoloration and rot develop at the root basal plate. Rotted areas of the bulb progress from the basal plate to the neck of the bulb. The stems and bulbs become reddish or reddish purple in color as the disease advances. The bulb tissues appear brown and watery when cut open. White fungal growth may develop at the base of the infected bulbs and stem plates. Roots turn dark brown in color and eventually rot and die. Severely affected plants come off easily when pulled slightly leaving the rotted basal plate and roots behind.

The pathogen is both soil-borne and seed-borne. The chlamydospores produced by the fungus can remain in the debris and in the soil indefinitely may cause primary infection. The fungus present in the garlic cloves used as seed material may also perpetuate the disease. The fungus is a typical wound parasite and enter into the plants through wounds or injuries caused during cultural operations or by nematodes or insects.

Moderately high temperature between 25° - 29°C and high soil moisture favor the occurrence and spread of the disease.

Control measures

- Avoiding *Fusarium* sick soil for raising garlic crop, deep summer ploughing before planting garlic, use of disease free seed material, removing and destroying crop debris from the field, following proper irrigation practices, following crop rotation practices, removing severely affected plants, avoiding injuries or wounds to plants and bulbs during cultural operations, eliminating alternate hosts from garlic fields, controlling nematode and other insect pests from causing injuries to garlic crop are measures that can be taken to prevent occurrence and spread of the disease.

General characters of genus *Fusarium* - Refer page - 138

xi) Rust disease of garlic - *Puccinia allii / Puccinia porri*

'Rust' disease of garlic is considered to be a minor one and occurs sporadically. Initial symptoms of the disease appear as scattered, small flecks, 1.0 - 2.0 mm in size, which enlarge into slightly larger, oval to diamond-shaped, reddish to orange colored pustules of size 3.0 - 5.0 mm on the leaf blades. Lesions develop first on older leaves and subsequently spread to younger leaves also. Reddish, air-borne urediniospores are produced in abundance in the lesions. Later these lesions may turn dark as resting spores (teliospores) are formed within the pustules. Severely affected leaves are completely covered with pustules and the leaves become chlorotic, dry out and die. Under severe infection of the disease, the plants produce small sized, poor quality bulbs.

Rust disease of garlic

The fungus overwinters on self sown, voluntary garlic plants as well as on alternate and collateral hosts, such as onion, shallots, leek, wild species of onion etc. The urediniospores produced from such hosts serve to initiate the disease in garlic.

The disease is more serious during cool weather conditions, when the temperature is between 14° - 24°C and when the humidity is very high near about 100% relative humidity.

Control measures

- Field sanitation and elimination of alternate hosts from garlic fields and near about garlic fields, crop rotation etc. help to prevent disease occurrence and spread.

- Fungicidal spraying with wettable sulphur 0.4% or mancozeb 0.2% helps to control the disease to some extent.

General characters of genus *Puccinia* - Refer page - 157

xii) Iris yellow spot disease of garlic - Iris yellow spot virus

'Iris yellow spot' disease is a minor virus disease. The affected plants are not killed but the yield and quality of the bulbs are adversely affected. The symptoms appear as straw-colored or creamy, elliptical or spindle-shaped spots without well defined margins on the leaves. These spots coalesce to form larger patches on leaves. The symptoms are more clear on older leaves. Similar spots appear on garlic scapes or flower stalks. In advanced stages of the disease, infected leaves and scapes collapse at the site of the spots.

Iris yellow spot of garlic

The onion thrips *(Thrips tabaci)* transmit the disease in a persistent manner. Disease intensity increases with the increase in population of the vector. The virus is not transmitted through seed. Overwintering infected garlic in the field, self sown volunteer plants, infected transplants and alternate hosts serve as sources of both vector and virus survival and disease spread.

Control measures

- Removal and destruction of plant debris, volunteer plants and alternate hosts, use of disease-free transplants etc. help to ward off the disease occurrence.
- The disease transmitting vector can be eradicated by use of suitable insecticides.

xiii) Bacterial brown rot or Slippery skin disease of garlic - *Pseudomonas gladioli* p.v. *allicola*

The plants may be affected at any growth stage of the crop. One or two leaves near the centre of the affected plant wilt, turn yellow and die back from the tips downwards while the other leaves appear healthy. During the early phases of the disease the bulbs may appear healthy, but the tissues in the neck region become soft and pulpy. One or more inner scales become water-soaked. The disease progresses from the top of the infected scales to the base. Eventually the entire internal tissues rot and turn brown. The internal scales dry and the bulb shrivel. When the base of the infected plant is squeezed, the rotted inner portion of the bulb slips out through the neck and hence the disease is also called as 'slippery skin'.

The bacteria are soil-borne and can survive in the debris in the soil as a saprophyte. The water-splashed bacteria reach the foliage and neck and enter the plant through wounds or injuries caused inadvertently.

High temperatures and high atmospheric moisture favor disease occurrence. Copious irrigation, water stagnation and persistent dew predispose the plants to infection.

Control measures

- Field sanitation, proper water management, removal and destruction of infected clumps, avoiding crop damage during cultural operations or by insect pests etc. are measures that can be adopted to curtail disease occurrence.

xiv) Powdery mildew of garlic - *Leveillula taurica*

'Leveillula taurica' is a serious parasite of garlic and causes considerable damage to the crop and yield loss. Symptoms of the disease appear as circular or oblong lesions, which appear chlorotic or necrotic. The lesions usually appear on older leaves before the bulbs begin to from. However, younger leaves are also attacked towards the end of the season. Severely affected leaves dry & die prematurely.

Powdery mildew of garlic

Cleistothecia, the resting spores found in crop residues on the soil are responsible for causing initial infection. Under favorable climatic conditions they release ascospores after the cleistothecial wall is ruptured. The ascospores, which are carried by wind or rain splash cause primary infection. The conidia produced in large numbers from conidiophores, which are carried by wind cause secondary infection.

Warm and dry weather conditions favor disease occurrence and spread.

Control measures

- Removal and destruction of infected plant debris from the soil, adopting crop rotation practices, removal of alternate weed hosts near the vicinity of garlic crop and following such other sanitary measures help to prevent the occurrence of the disease.

- Spraying the crop with Propiconazole 0.15% or Hexaconazole 0.15% or Dinocap 0.1% or Wettable sulphur 0.4% has been found to be effective in controlling the disease. The first spraying is given as soon as initial symptoms appear, followed by 2 - 3 sprayings at fortnightly intervals.

General characters of Genus *Leveillula*

7. Coriander *(Coriandrum sativum)*

'Coriander' is a small, flowering, annual herb in the Family-Apiaceae. It is also known as 'Chinese Parsley'. It is probably native to the Middle East and Southern Europe, but it has been grown in Asia and the Orient for centuries. Now it is cultivated in almost all the countries of the world. Coriander is found wild in Egypt.

Coriander is an important spice crop having a prime position in flavoring food. Its leaves and fruits have a unique and pleasant aroma and are commonly used raw or dried for several culinary applications including soups, chutneys etc. Coriander is used as a flavoring agent in several medicines, tobacco products and in cosmetics and soaps to give fragrance.

Coriander plants

Coriander seeds

Coriander has many medicinal properties. It is used for digestive problems including stomach upset, loss of appetite, nausea, diarrhea, bowel spasms and intestinal gas. It is also used to treat hernia, measles, hemorrhoids, tooth aches, worms, joint pains as well as many other infection caused by bacteria and fungi. Coriander is good for diabetes patients, as it stimulates insulin secretion. It also helps to lower blood sugar. The antioxidants present in coriander help in preventing eye diseases.

Diseases of coriander (*Coriandrum sativum*)

i) Powdery mildew of coriander - *Erysiphe polygoni*

Usually the disease appears during the later stages of crop growth. Initially small, whitish, circular to irregular, powdery patches appear on both the surfaces of leaves. Soon the disease spreads to all green parts of the plants. The powdery patches enlarge rapidly and often cover the entire area of the leaves, young branches, petioles and the umbels. The plants appear as if covered by a grayish-white, powdery coating from a distance. The leaves are also reduced in size and are distorted. Eventually the leaves turn yellow, wither and dry. Sometimes the seeds are also attacked. The whitish powdery coating comprises of conidiophores and conidia produced from the superficial mycelium. Numerous minute, black, pin point-like cleistothecia are formed later on, which are embedded in the mycelial weft. Infected plants become very much stunted

Powdery mildew of coriander

The pathogen perennates through cleistothecia in the infected plant debris in the soil. The ascospores released from the cleistothecia infect the lowermost leaves near the soil first. Secondary spread occurs by means of conidia produced in profusion from the mycelium on the infected leaves. The pathogen has many other alternate hosts, such as cabbage, sugar beet, turnip, clover, lentil, *Vicia, Trifolium, Lupinus* etc., and so there is a steady supply of conidia from these hosts, which are carried by wind currents to coriander and cause fresh infection.

The disease occurrence is favored by dry conditions. Cool nights and warm days are favorable for disease occurrence and spread. Temperatures between 20° - 24°C and relative humidity up to 70% are conducive for the development of the disease.

Control measures

- Removal and destruction of infected plant debris from the soil, adopting crop rotation practices, removal of alternate weed hosts near the vicinity of coriander crop and following such other sanitary measures help to prevent the occurrence of the disease.
- Spraying the crop with Propiconazole 0.15% or Hexaconazole 0.15% or Dinocap 0.1% or Wettable sulphur 0.4% has been found to be effective in controlling the disease. The first spraying is given as soon as initial symptoms appear, followed by 2 - 3 sprayings at fortnightly intervals.

General characters of genus *Erysiphe* - Refer page - 159

ii) Stem gall of coriander - *Protomyces macrosporus*

'Stem gall' is the most important disease of coriander and causes considerable yield loss. The symptoms appear as galls, which develop as chemically induced swellings arising from the surface of the leaf lamina, veins, mid ribs and petioles. On the leaf lamina the galls appear as yellow bulges on the upper surface. On veins, mid ribs and petioles, the galls appear as translucent, yellowish-white swellings that are often elongated and blister-like. The galls, which are soft and fleshy when young becomes hard and woody later on. The galls rupture and release ascospores. The peduncles are malformed and the flowers and fruits are hypertrophied. In case of severe attack the plants may be killed.

The chlamydospores present in the plant debris in the soil are responsible for causing initial infection. The disease is also carried through chlamydospores present in the seeds. The ascospores produced in the asci and released in large numbers from the ruptured galls cause secondary infection.

High soil moisture, low night temperature and dew fall favor the disease occurrence and spread. The disease severity is more in shady places.

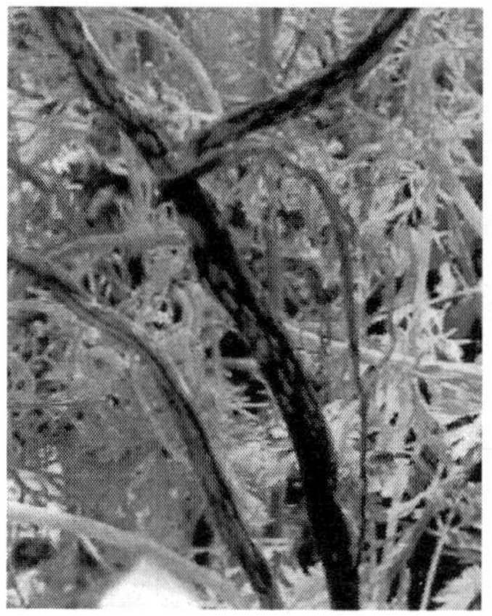

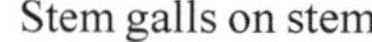

Stem galls on stem

Stem galls on flowers and fruits

Control measures

- Use of contamination free, healthy seeds, field sanitation, provision of adequate drainage facilities and crop rotation practices help to reduce the inoculum potential.
- Seed treatment with Thiram at 4.0 g / Kg or carbendazim at 2.0 g / Kg of seeds is effective in eradicating the externally seed-borne inoculum.
- Spraying the crop with mancozeb 0.2% or captafol 0.15% or carboxin 0.1% or carbendazim 0.1% has been found to be effective in controlling the disease.

General characters of genus *Protomyces* - Refer page - 160

iii) *Fusarium* wilt of coriander - *Fusarium oxysporum* f.sp. *corianderii*

Wilt of coriander

The pathogen attacks the seedlings before emergence as well as grown up plants. Early infection occurs on the radical of seedlings, which causes the seedlings to wilt, collapse and die. The symptoms appear like the symptoms caused by damping off disease. When older plants are infected they exhibit stunting and yellowing symptoms. The lesions caused in the underground hypocotyl of the infected plants coalesce and girdle the stem region as a result the plants collapse and die. Vascular tissues of infected older plants show red to brown discoloration of the xylem vessel. The roots are also affected by the pathogen and they rot and die.

The pathogen is soil-borne and overwinters as chlamydospores. They germinate under favorable conditions, produce mycelium and causes primary infection. The mycelium, which is found around the stems of infected plants near the soil level produce both micro-and macro conidia in profusion and cause secondary infection.

Moderate temperature of about 28°C, high soil moisture and slightly acidic nature of the soil with pH 5.8 - 6.9 favor disease occurrence and spread.

Control measures

- Field sanitation, adoption of crop rotation practices, avoiding raising of coriander crop in already contaminated fields etc. are measures that can be taken to prevent occurrence of the disease.
- Seed treatment with carbendazim at 2.0 g / Kg of seed helps to ward off the disease in the early stages.
- Seed treatment with *Trichoderma viride* at 4.0 g / Kg or *Pseudomonas fluorescens* at 10 g / Kg of seed has been found to afford control of the disease in the seedling stage.
- Soil application of *Trichoderma viride* at 10 Kg / acre or *Pseudomonas fluorescens* at 2.0 Kg / acre has also been found to be effective in controlling the disease.

General characters of genus *Fusarium*-Refer page - 138

iv) Bacterial leaf spot of coriander-*Pseudomonas syringae* pv. *coriandricola*

Initial symptoms appear as small, water-soaked spots between leaf veins on both surfaces of leaves. The spots enlarge and turn dark brown to black. The spots are mostly angular and are limited by the veins. Severe incidence of the disease leads to blighting of large areas of the leaves. The bacteria enter the petioles and stems and attack the vascular elements and cause elongated dark streaks. The inflorescences become yellow, which later turn brown and are blighted. Flowers and seed heads shrivel, turn black and collapse. Water-soaked lesions develop on the fruits also.

Bacterial leaf spot of coriander

The disease is transmitted through infected seed. The bacteria can remain in the plant debris in the soil and are spread by splashing irrigation water and rain.

Some insects also carry the bacteria externally and spread the disease.

Control measures

- Using disease-free seeds for sowing, removal and destruction of debris from the soil, avoiding splash irrigation, removal and destruction of severely affected plants, following crop rotation practices and such field sanitation measures can be adopted to prevent occurrence of the disease.
- Spraying the crop with Agrimycin 100 (0.2 - 0.4 g / litre of water) when initial symptoms of the disease appear, followed by 2 - 3 more sprayings has been found to be effective in controlling the disease. Combined spraying with Agrimycin 100 and copper oxychloride 0.25% has been found to afford better control of the disease.

v) Damping off disease of coriander-*Fusarium* spp., *Pythium* sp. and *Rhizoctonia solani*

The disease occurs often in coriander nurseries. In the first phase of the disease, the seeds are affected and they become soft, rot and fail to germinate. In the second pre emergence phase, the seedlings die rapidly before they emerge out from the soil. In the third post emergence phase, the seedlings collapse and die in large numbers. After emergence from the soil, the stems of the seedlings become water-soaked and reddish-brown lesions develop and girdle the stem and rot at the soil level. The affected seedlings topple and die in patches.

The pathogens are soil-borne and can exist as saprophytes in the debris in the soil indefinitely and initiate infection. The resting spores produced by the organisms and the mycelium are carried by running water and by equipments and spread the disease.

High soil moisture and moderate to high temperatures favor the disease occurrence and spread

Control measures

- Removal and destruction of debris from the soil, providing proper drainage facilities, judicious water management, avoiding sick soils for raising nurseries, avoiding thick sowing etc. are measures that can be followed to prevent occurrence of the disease.
- Seed treatment with thiram or captan at 4.0 g / Kg of seed and drenching the soil with Bordeaux mixture 1.0% or copper oxychloride 0.25% or thiram or captan 0.2% afford good control of the disease.

vi) Stem rot of coriander - *Sclerotinia sclerotiorum*

The disease appears mostly at the time of flowering. Initially small, water-soaked spots appear on the main stem which turn brown. The spots enlarge and the whole stem near about the soil level is completely girdled. The affected areas are covered with a white, fluffy mycelial growth. Sometimes the outer skin of the stem is ruptured. When the xylem vessels are attacked, movement of nutrients and water to the aerial parts of the plants are restricted, as a result the plants wilt and die. Minute, black sclerotia are produced in large numbers, which are embedded in the mycelium and on the surface of the affected stem.

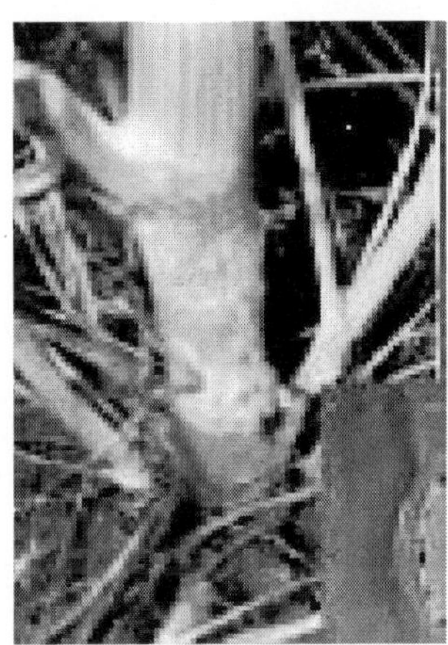

Stem rot of coriander

The disease is mainly soil-borne. The sclerotia can remain in a viable state for indefinite periods in the debris in the soil. The sclerotia are carried to other fields and places through water and by equipments and cause primary infection. The ascospores produced in large numbers from asci in the apothecia cause secondary infection.

Wet conditions, high soil moisture and a temperature range from 15° - 21°C are conducive for the occurrence and spread of the disease.

Control measures

- Removal and destruction of debris from the soil, using sclerotium free seeds, judicious water management practices, removal of disease affected plants, crop rotation, elimination of alternate hosts are measures that can be followed to prevent occurrence of the disease.
- Soil drenching with mancozeb 0.2% or copper oxychloride 0.25% or captan 0.2% or thiram 0.2% when initial symptoms of the disease appear is effective in eradicating the inoculum in the soil.

- Foliar application of chlorothalonil o.15% or thiophanate methyl 0.1% or iprodione 0.15% is recommended for the control of the disease.

General characters of genus *Sclerotinia* - Refer page - 161

vii) Leaf spot of coriander - *Alternaria alternata*

Leaf spot of coriander

The disease mostly attacks the leaves. Symptoms of the disease appear as small, scattered, dark brown to almost black, circular spots The spots gradually enlarge and become larger spots, which are irregularly circular, dark brown in color with a light brown margin and surrounded by a chlorotic or yellow halo. Later the centre of the spots turn grayish-white and necrotic. The spots may show concentric rings. When the disease incidence is severe the spots coalesce and cover large areas of the leaves resulting in blighting and defoliation. The pathogen may attack the leaf petioles and young stems and produce linear, necrotic lesions.

The pathogen can remain in the diseased plant debris in the soil for months together and continue to produce conidia. The conidia are carried through air & cause fresh infection. The pathogen has got many alternate hosts and the conidia produced from such hosts may cause infection in coriander crop any time.

Control measures

- Field sanitation, clean cultivation, crop rotation etc. may help to minimize disease occurrence.
- Seed treatment with captan or thiram at 4 g / Kg of seeds protects the young plants from infection.
- Spraying the crop with copper oxychloride 0.25% or mancozeb 0.2% or ziram 0.2% when initial symptoms of the disease appear, followed by 2 - 3 more sprayings at fortnightly intervals affords good control of the disease.
- Foliar spraying with hexaconazole (Contaf) 0.1% or propiconazole (Tilt) 0.1% has been found to be very effective.

General characters of genus *Alternaria*-Refer page - 153

viii) Anthracnose of coriander-*Colletotrichum gloeosporiodes* and *C. capsici*

Symptoms appear as small, scattered brown to dark brown spots on the leaves. The spots enlarge and turn into irregularly circular necrotic lesions with dark margin and ashy gray centre. Necrotic lesions develop on the stem region also. Severely affected leaves may drop off prematurely. On the lesions minute, brown to black , pin point-like dots, which are the acervuli of the pathogen are formed in large numbers. Seedlings, which are 5 - 6 weeks old are most vulnerable to attack by the pathogen and hence the disease is also called as **'Seedling blight'**.

The disease is both seed-borne and soil-borne. Diseased plant debris in the soil is the main source of perennation of the disease. The conidia produced from such sources initiate fresh infection. Secondary spread is by air-borne conidia The pathogens have several alternate hosts and the conidia produced from such hosts are also responsible for the spread of the disease.

High relative humidity of 95 - 97%, misty conditions and moderate temperatures of about 25°C are conducive for the occurrence and development of the disease.

Control measures

- Removal and destruction of diseased fallen leaves and debris from the soil, elimination of alternate hosts from coriander fields and nearby areas, adopting crop rotation practices etc. can help in preventing the occurrence of the disease.
- Seed treatment with captan or thiram at 4.0g / Kg of seed eliminates the external seed-borne inoculum.
- Spraying with Bordeaux mixture 1.0% or copper oxychloride 0.25% or mancozeb 0.2% or carbendazim 0.1% ot thiopanate methyl 0.1% or chlorothalonil 0.15% affords good control of the disease. Spraying has to be given once the initial disease symptoms appear followed by 2 - 3 sprayings at fortnightly intervals.

General characters of genus *Colletotrichum* - Refer page - 136

ix) Grain mould of coriander - *Fusarium* sp., *Curvularia* sp., *Alternaria* sp.

Several species of fungi are involved in causing grain mould of coriander. The infection starts after the plants start flowering. The flowers at the time of seed setting are first affected and the infection continues till seed maturity. Due to infection and colonization of the mould fungi, the grains at various stages of development are discolored. Grain discoloration varies from whitish, pinkish, grayish to blackish depending upon the fungi colonizing the grains. In severe cases, the grains turn completely black. When infection starts early, grain development is adversely affected and the grains produced are shrivelled and chaffy. When conditions are favorable for the organisms causing mould, the yield and quality of the grains are reduced markedly.

The disease is caused by air-borne fungal spores, which float in the air throughout the year and infect the grains when conditions turn favorable for them.

Highly humid conditions with relative humidity above 90%, frequent rains and high temperatures ranging from 25° - 35°C during flowering to grain development period are favorable for disease occurrence.

Control measures

- Spraying the aerial parts with Auriofungin Sol 200 ppm. or captan 0.2% or carbendazim 0.1% from the time of flowering has been recommended for controlling the disease. When conditions are favorable for the pathogens, if rain continues repeated sprayings may be necessary.

8. Fenugreek *(Trigonella foenum - graecum)*

'Fenugreek' (Methi) is an annual, herbaceous, flowering legume belonging to Family Fabaceae. It grows well in the semi arid regions. It is an all purpose, aromatic, culinary herb. Fenugreek is used as a spice throughout the world to enhance the sensory quality of various foods. Seeds and leaves are common ingredients in many culinary dishes in the Indian sub continent. Fenugreek has also got many medicinal uses. The seeds have a characteristic strong and pleasant odor. The leaves are also used as a green.

Fenugreek is native to South Eastern Europe, Western Asia and the Indian subcontinent. India occupies a dominant position in world production and export of fenugreek. Other major fenugreek growing countries include North Africa, Mediterranean Europe, China, South East Asia, Australia, United States of America, Argentina, France and Canada. Within India Rajasthan is the largest producer of fenugreek followed by Gujarat, Madhya Pradesh, Tamil Nadu, Punjab and Uttar Pradesh.

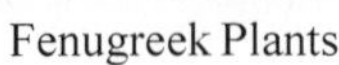

Fenugreek Plants

Fenugreek Seeds

The seeds and leaves of fenugreek are edible. The ground seeds are often used in curry powders and spice blends. Fenugreek serves as an integral ingredient in Indian cuisines, especially in curries, chutneys, pickles and many other vegetarian and non-vegetarian dishes. Addition of fenugreek gives a distinctive, sweet flavor resembling that of maple syrup to the food items. The chemical 'stoloton' infuses the aroma in the food items. Because of the maple syrup like flavor, it is added to ice creams, beverages, soups, soaps and cosmetics. The leaves are used in salads and both fresh and dried leaves are used in Indian cookery.

Besides culinary values, fenugreek has several medicinal values. Fenugreek seeds encourage weight loss; help in fighting against all digestive problems, acid refluxes and heart burns by flushing out harmful toxins from the body; reduce cholesterol level; improve diabetes symptoms by producing more insulin and lower the blood sugar level; reduce menstrual cramps, improve lactation in breast feeding mothers; reduce Arthritis pain; improve kidney functioning; prevents hair loss; promote hair growth and prevent premature grey hair, cleans the skin; reduce blemishes and dark circles in the skin.

Drinking fenugreek water daily can provide relief to many of these problems. Fenugreek water is prepared by soaking two table spoons of fenugreek seeds in two glasses of water overnight. The water is drained in the next morning and the health drink is consumed on an empty stomach. Some of the soaked fenugreek seeds may also be eaten.

Diseases of fenugreek (*Trigonella foenum - graecum*)

i) *Cercospora* leaf spot of fenugreek - *Cercospora traversiana*

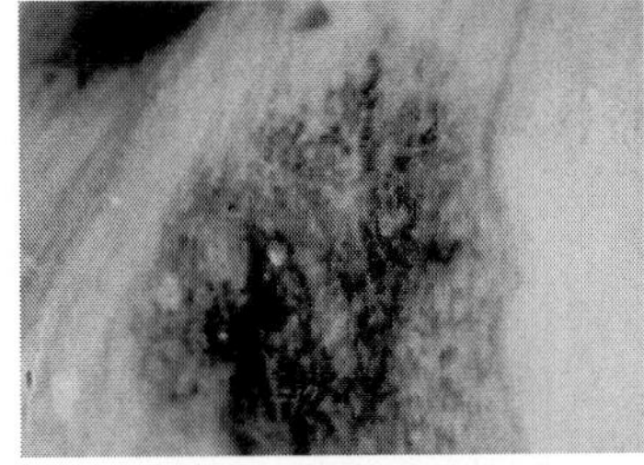
Cercospora leaf spot of fenugreek

It is one of the most serious diseases of fenugreek and causes considerable damage to the crop and yield loss. Symptoms of the disease appear first on older leaves and then progresses to the younger leaves. Symptoms appear on the leaves as sunken, gray colored to dark tan colored lesions with chlorotic halos. The lesions coalesce and cover large areas of the leaves and become necrotic. Severely affected leaves wither and die prematurely. From the substomatal cavities in the grayish tan lesions black dots, which are the pseudostromata of the pathogen appear. The pseudostromata produce conidiophores in clusters and conidia are borne on the conidiophores. Under conditions of high humidity, the entire lesions appear fuzzy due to the presence of numerous conidia. Necrotic patches may develop on the pods also.

The pathogen survives mainly in the plant debris as pseudostromata, which can resist desiccation and produce conidia continuously when moist conditions prevail. The conidia can also survive in the debris in the soil and seeds of the host. The conidia produced are spread by rain splash, wind and by other agencies and cause secondary infection.

Moderate to high day temperatures ranging from 25° - 35°C., and night temperature above 16°C. and high relative humidity of 90 - 95% and long period of leaf wetness favor disease development and spread. When the temperature goes down below 15°C. and when there is no leaf wetness for less than 11 hours the disease is absent completely.

Control measures

- Removal and destruction of plant debris from the soil, adopting crop rotation practices, providing adequate drainage facilities and such field sanitation measures help to ward off the disease occurrence and spread.
- Foliar spraying with Bordeaux mixture 1.0% or copper oxychloride 0.25% or mancozeb 0.2% or chlorothalonil 0.15% or carbendazim 0.1% when initial symptoms of the disease appear followed by 2 - 3 sprayings at fortnightly intervals controls the disease.

General characters of genus *Cercospora* - Refer page - 144

ii) Powdery mildew of fenugreek - *Erysiphe poligoni*

Powdery mildew of fenugreek

'Powdery mildew' of fenugreek is more severe than downy mildew. Initial symptoms appear as small, irregular, powdery patches on the upper surface of leaves. These powdery patches enlarge and may cover the entire leaf area. White powdery patches appear on the petioles,stems and the pods and the plants appear grayish-white from a distance. The affected leaves turn yellow, become chlorotic and drop off prematurely. The white powdery coating comprises of conidiophores and conidia, which are produced abundantly. The fungus may infect the young seeds inside the pods also. Later in the season, minute, black cleistothecia, which are the sexual fruiting bodies of the pathogen are formed embedded in the mycelial weft.

The disease perennates through cleistothecia in the plant debris in the soil. The asospores liberated from the asci after disintegration of the cleistothecial wall infect the lower most leaves near the also survive as ground level and cause primary infection. The pathogen can also survive as dormant mycelium on infected plant parts, as well as in infected seeds. Secondary infection occurs by air-borne conidia.

Warm and humid climate favor the occurrence and spread of the disease. Moderate temperatures about 25°C and dry conditions are quite conducive for the disease occurrence and spread.

Control measures

- Collection and destruction of debris from the soil, adopting crop rotation practices, field sanitation etc. are agronomic practices that can be adopted to ward off the disease occurrence.
- Seed treatment with thirm or captan at 4,0 g / Kg of seeds is effective in eliminating externally seed-borne inoculum.
- Spraying with wettable sulphur 0.4% or carboxin 0.1% or dinacap 0.15% at the onset of the disease followed by a second spraying 15 days later is effective in controlling the disease.

General characters of genus *Erysiphe* - Refer page - 159

iii) Powdery mildew of fenugreek - *Leveillula (Oidiopsis) taurica*

Fenugreek is subjected to attack by another powdery mildew disease caused by '*Leveillula taurica*'. The symptoms caused by both the pathogens are almost similar. In the case of *Leveillula*, the whitish, powdery coating consisting of conodiophores and conidia appear mostly on the upper surface of leaves, while the corresponding undersurface of leaves turn yellow. The fungus soon covers the entire leaf lamina and the leaves turn completely yellow, dry and drop off prematurely.

Warm and humid conditions favor disease development and spread. High relative humidity favor conidial production and low relative humidity favor spore dispersal and spread of the disease.

The measures suggested for the control of powdery mildew caused by *Erysiphe poligoni* is applicable for the control of this powdery mildew disease also.

General characters of genus *Leveillula* - Refer page - 152

iv) Charcoal rot of fenugreek - *Macrophomina phaseolina*

'Charcoal rot' also known as 'Dry weather wilt' usually occurs late in the crop season during the reproductive stages. Infected plants produce slightly smaller leaves with reduced vigor. Gradually the leaves turn yellow, wilt, turn brown and die. But the dead leaves remain attached to the leaf petioles. The stem at the soil level and the tap root turn brown. A light gray or silvery discoloration is visible in the tap root and lower stem when split open. Minute, black, dot-like structures, which are the microsclerotia produced by the pathogen are visible in the tissues of the stem and tap root. Outer tissues are covered with masses of black, dusty microsclerotia presenting a characteristic charcoal-like appearance. Seed setting is adversely affected in the affected plants.

Hot, dry weather conditions favor disease occurrence. The mycelium from the germinating microsclerotia cause initial infection. The pathogen is soil-borne and survives as microsclerotia in the soil and in infected plant debris for a period of 3 years or more under dry conditions.

Control measures

- Removal and destruction of plant debris in the soil, field sanitation, removal and destruction of severely affected plants, proper maintenance of the crop so as to reduce any stress conditions to the crop, adoption of crop rotation practices are measures that can be taken to ward off the disease.

General characters of genus *Macrophomina* - Refer page

v) Root rot / Collar rot / Foot rot of fenugreek - *Rhizoctonia solani* and *Sclerotium rolfsii*

Primary infection is initiated by the mycelium arising from the sclerotia. Infection occurs from the growing tips of young lateral roots. The pathogen progresses from the root tips to the main root resulting in rotting of the main root. *Rhizoctonia* affected rootlets often develop characteristic 'spear-point' symptom. Symptoms on the leaves appear as small, scattered, chlorotic, water-soaked lesions on both the surfaces of leaves. The lesions enlarge and become irregular-shaped spots, which turn brown with dark brown margins. The central portion of these spots later become pale, dry and disintegrate. The affected plants are stunted. In the advanced stages the pathogen produces brown to black, globular to irregularly shaped sclerotia in large numbers in the affected tissues.

Foot rot of fenugreek

The pathogen is mainly soil-borne. The sclerotia survive in the diseased plant debris and in the soil for long periods in the absence of a host.

High humidity and warm weather conditions favor the development of the disease.

Control measures

- Use of uncontaminated seeds, removal and destruction of debris from the soil, removal and destruction of severely affected plants, proper maintenance of the crop, adopting crop rotation practice wherever possible, application of adequate quantities of farm yard manure or compost to the soil so as to encourage the growth of antagonistic microorganisms, which can destroy the pathogen are measures that can be followed to control the disease occurrence and spread.
- Seed treatment with the biological agent *Trichoderma viride* at 4.0 g / kg of seed and soil application of *Trichoderma viride* at 2.0 Kg / acre at the time of sowing is effective in controlling the disease.

vi) Downy mildew of fenugreek - *Peronospora trigonellae*

'Downy mildew' is considered to be a minor disease of fenugreek. The disease is more severe during the flowering stage of the crop. The symptoms appear as yellow patches or small chlorotic spots on the upper surface of the leaves mostly near about the leaf margins. On the corresponding undersurface of the patches fine, white, cottony downy growth develops, which appear grayish white. This weft of fungal growth comprises of mycelium, sporangiophores and sporangia. In old lesions oospores, which are the sexual spores are produced and are found embedded in the diseased leaf tissues.

The disease is primarily soil-borne. The oospores survive in the diseased plant parts and plant debris in the soil. They germinate by producing a germ tube and infect the plants and cause primary infection. Secondary infection is caused by wind-borne sporangia.

Low temperatures and high humidity favors disease occurrence and spread. Maximum temperature of 18° - 24°C and minimum temperatures of 4° - 10°C and relative humidity above 80% is conducive for disease development. Maximum infection occurs when there is leaf wetness for 12 hours.

Control measures

- Removal and destruction of diseased crop debris in the soil, removal of severely affected plants, adopting crop rotation practices help to ward off the disease.
- Spraying with mancozeb 0.2% as soon as initial symptoms appear helps to prevent further spread of the disease. Spraying with metalaxyl M 0.1% or mancozeb 0.2% + metalaxyl 0.1% is more effective in controlling the disease.

General characters of genus *Peronospora* - Refer page - 150

vii) Rust of fenugreek - *Uromyces trigonellae*

'Rust' of fenugreek is a minor disease and occurs sporadically and may cause extensive damage to the crop and yield loss when the incidence of the disease is severe. The disease mostly attacks the leaves, rarely the petioles and stems. The symptoms are more pronounced on the under surface of leaves, but may appear on the upper surface also. Initial symptoms appear as minute, whitish, slightly raised pustules. These rust pustules may enlarge up to 2.0 mm in diameter and become reddish-brown. These sori contain uredospores in large numbers. A ring of secondary sori may be formed around the primary sori. When the sori rupture by breaking the host epidermis, masses of powdery uredospores are released. Too many uredosori and masses of uredospores on the leaf surface give a rusty appearance to the leaves. Telia appear quite late in the season in the uredosori or separately. The teliosori appear dark brown or black. Severely affected leaves t urn yellow, dry and drop off prematurely.

The fungus survives mainly through teliospores, which are thick-walled resting spores present on the leaves left over in the field or on the soil surface. Secondary infection is through wind-borne uredospores, which are the main source of perpetuation of the disease. When conditions are favorable for the pathogen, it can complete the infection cycle within 5 days and a new crop of uredospores is produced within 5-10 days. Thus, the uredial stage constitute the repeating stage in the life cycle of the rust fungi. The disease may also spread through wind-borne aeciospores. The disease may also spread through spores produced in alternate hosts.

Low temperature ranging from 20°-26°C and high relative humidity of 86- 92% favor disease development. Cloudy, damp and humid day and dew during the early morning hours are favorable for spore germination and infection. The uredospores, which are the repeating spores, germinate best at 15° - 24°C

Control measures

- Removal and destruction of infected crop debris in the field, proper maintenance of the crop, adoption of crop rotation practices help to check disease occurrence.
- Spraying with wettable sulphur 0.4% or mancozeb 0.2% or chlorothalonil 0.15% or triadimefon 0.1% as soon as initial symptoms of the disease is observed followed by 2 - 3 more sprayings at fortnightly intervals controls the disease.

General characters of genus *Uromyces* - Refer page - 164

viii) Damping off of seedlings of fenugreek - *Pythium aphanidermatum*

Damping off of seedlings of fenugreek

'Damping off of seedlings' may occur as pre-emergence damping off and post-emergence damping off. In pre-emergence damping off, the seed and the radicle rot before the seedling emerges from the soil. In post-emergence damping off, which is more conspicuous, the newly emerged seedlings are killed after emergence from the soil. Infection starts at or below the ground level. The basal portion of the affected stem becomes soft, water-soaked and discolored due to the death of the cortical tissues. As the disease advances, the stem becomes constricted at the base, turns dark and the seedling topples over, rots and dies. The rotten seedlings emit a bad odor.

The disease is soil-borne. The pathogen is a natural inhabitant of soil and can survive on dead plant and animal wastes and refuse in the soil as a saprprophyte. The oospores produced by the pathogen, which remain dormant, germinate and cause primary infection. Secondary infection is caused by motile zoospores released from sporangia. The spores are spread mostly by rain splash and by wind.

High soil moisture, high relative humidity, dampness due to heavy rainfall and soil temperatures between 24° - 30°C, hard clay soils and crowded seedlings favor disease occurrence. Cloudiness and low temperatures below 24°C for a few days are favorable for infection and development of the disease.

Control measures

- Raising nurseries in elevated places with good drainage facilities, avoiding thick sowing in the nurseries, avoiding raising nurseries in sick soils, avoiding excessive irrigation and following field sanitation measures help to prevent the occurrence of the disease to a large extent.
- Seed treatment with captan or thiram at 4.0 g or metalaxyl 2.0 g / Kg of seeds provide protection to the seedlings.
- After sowing the soil may be drenched with Bordeaux mixture 1.0% or captan 0.2% or thiram 0.2% or copper oxychloride 0.25% followed by a second drenching after emergence of the seedlings
- Seed treatment with *Trichoderma harzianum* at 4.0 gm / Kg of seeds or *Pseudomonas fluorescens* at 10 g / Kg of seeds protects the seedlings from initial infection in the nursery.

General characters of genus *Pythium* - Refer page - 133

ix) *Fusarium* wilt of fenugreeek - *Fusarium oxysporum*

Fusarium wilt of fenugreek

'***Fusarium* wilt'** may attack both young as well as mature plants. Initial symptoms appear in young plants as minor vein clearing of young leaflets accompanied by downward drooping of older leaves and the affected plants may wilt and die soon. In mature plants vein clearing, downward drooping of leaves, stunting of plants, yellowing of the lower leaves, wilting of leaves and young stems are seen. This is followed by marginal necrosis of the infected leaves, rapid defoliation and death of the plants. Vascular tissues show brown discoloration. The symptoms are more prominent in older plants at flowering and pod formation stages.

The pathogen is mainly soil-borne and can survive as a saprophyte for several years. The sclerotia formed by the pathogen can also remain viable in the soil for several years and cause primary infection. The pathogen is also seed-borne to some extent as the sclerotia may be carried on the seed surface. The spores of the pathogen are disseminated through soil, infected plant debris and seed. The pathogen has several alternate hosts.

Soil temperatures of 24° - 28°C and soil moisture levels of 80 - 90% are favorable for the occurrence and spread of the disease. Temperatures above 35°C inhibit the development of the disease. The disease occurs more in heavy clay soils.

Control measures

- Deep summer ploughing so as to expose the mycelium and spores to the soil surface and their destruction by direct exposure to sun light, removal and destruction of plant debris from the soil, removal and destruction of severely affected plants, application of farm yard manure or compost at the rate of 5.0 tons per acre so as to encourage the growth of actinomycetes and other antagonistic microorganisms, which may destroy the pathogen, long term crop rotation practices and proper maintenance of the crop by applying macro and micro nutrients as per recommendations are measures that can be followed to control the disease.
- Seed treatment with captan or thiram at 4.0 g / Kg of seed is effective in eliminating the pathogen on the seeds.

General characters of genus *Fusarium*-Refer page - 138

x) Yellow mosaic disease of fenugreek-Bean yellow mosaic virus

'Yellow mosaic disease' of fenugreek is a minor disease and occurs sporadically. The virus causing the disease belongs to 'Potyvirus'. The symptoms of the disease are vein clearing, severe mottling of leaves, curling of the leaves at the margins and reduced leaf size. Pods are usually underdeveloped and mottled. Seeds are thin and shrunken.

The virus is transmitted by aphids in a semi persistent manner. More than 20 species of aphids, especially *Acyrthosiphum pisum, Myzus persicae, Aphis fabae* & *Macrosiphum euphorbiae* are involved in transmitting the virus. The incidence of the disease is more when the population of the aphids are more. Both the virus as well as the vectors have many alternate hosts, as such the virus can be transmitted to the crop any time & at any growth stage of the crop.

Aphids are more active during warm periods. Under conditions favorable for the multiplication of the aphids they multiply in large numbers and spread the disease. The disease can be controlled to some extent by controlling the aphid vectors by application of suitable insecticides.

9. Cumin *(Cuminum cyminum)*

'Cumin' commonly known as **'Jeera"** is considered to be a native of Eastern Mediterranean and South Asia. Now it is cultivated in many countries of the world especially in India, Iran, Indonesia, China and Mexico. India is the largest producer of cumin in the world. In India, Gujarat and Rajasthan are the leading producers of cumin and contributes 70% of the total production.

Cumin plant belongs to Family - Apiaceae (Umbelliferae). It is an annual, herbaceous, glabrous, flowering plant that grows up to a height of 30 - 50 cm. The leaves are finely dissected and the flowers are white or pink colored and are borne on compound umbels. The seeds are oblong in shape, yellowish-brown in color and are longitudinally ridged.

Cumin Plants

Cumin Seeds

Cumin is grown for its pleasantly aromatic seeds. Cumin seeds are commonly used as a spice in Latin American, Middle Eastern, African and Indian cousins both as whole seed and as ground powder. Whole cumin seeds are used in many bakery products and in some cheeses. Ground cumin is an essential spice in curry powder, garam masala and in many savory spice mixtures.

Cumin has many health benefits too. It promotes digestion by increasing the activity of digestive enzymes. Cumin is very rich in iron and can supply the much needed iron and correct iron deficiency. Cumin contains many plant compounds, such as terpenes, phenols, flavonoids and alkaloids, which function as antioxidants and provide relief from heart ailments, diabetes, blood sugar, blood cholesterol and promote weight loss. Regular use of cumin in the daily diet is enough to give such health benefits.

Diseases of cumin (*Cuminum cyminum*)

i) Wilt disease of cumin - *Fusarium oxysporum* f.sp. *cumini*

'Wilt disease' of cumin attacks the crop at all growth stages. However the disease is more severe in younger plants. Infected plants show peculiar symptoms of drooping of tips of leaves, followed by wilting and death of the entire plants. The roots of affected plants show brown discoloration. When the stem of an infected plant is split longitudinally brown discoloration of the vascular bundles is seen. Diseased plants produce only a few, small, thin, shriveled seeds.

Fusarium wilt of cumin

The fungus is both seed-borne and soil-borne. After the death of infected plants, the fungus invades all the tissues, sporulates and infects neighboring plants. The fungal mycelium can survive as saprophyte in the diseased plant debris. The chlamydospores produced by the fungus can remain in a viable state in the soil for a number of years and cause fresh infection. The fungus spreads through irrigation water, rain splash, wind and through intercultural operations as well as by cattle.

High soil moisture and low soil temperatures between 12.5° - 14°C favor disease occurrence and spread of the disease.

Control measures

- Avoiding raising cumin crop in sick soils, deep summer ploughing so as to expose the mycelium and spores to the hot sun, removing and destroying diseased plant debris from the soil, use of disease-free, healthy seeds, removing and destroying severely affected plants, adopting crop rotation practices, providing adequate drainage facilities are measures that can be followed to avoid disease occurrence.

- Seed treatment with captan or thiram at 4.0 g / Kg of seed or carbendazim at 2.0 g / Kg of seed affords protection to seedlings from initial infection.
- Seed treatment with *Trichoderma harzianum* at 4.0 g / Kg of seed eliminates the pathogen and protects the seedlings from primary infection.
- Application of neem seed cake at 2.0 tons / acre in the field before planting helps to encourage the growth of antagonistic microorganisms in the soil that can destroy the fungal pathogen and prevent disease occurrence.

General characters of genus *Fusarium* - Refer page - 138

ii) Powdery mildew of cumin - *Erysiphe polygoni*

Powdery mildew of cumin

The disease occurs mostly during the flowering stage of the crop. Initial symptoms appear as small, irregular, powdery patches on the upper surface of leaves. These powdery patches enlarge and may cover the entire leaf area. Both the surfaces of the leaves and all the succulent parts, such as petioles, stems, branches and even the flowers are soon covered by the white powdery coating. The affected plants appear grayish-white from a distance. Infected leaves gradually turn yellow completely, die and fall off prematurely. The white powdery, chalky coating consists of fungal mycelium, conidiophores and conidia. When the production of conidia is almost over, numerous, small, black cleistothecia, which are the sexual fruiting bodies are produced and are embedded in the mycelial mat.

The pathogen perennates as cleistothecia in the diseased plant debris in the soil. The ascospores liberated from the asci inside the cleistothecia infect the lowermost leaves near the soil first and then the disease spreads to the other parts of the plant. Secondary spread of the disease is through conidia produced abundantly by wind and rain splash.

Cool nights and warm days are favorable for the development of the disease. Moderate temperatures of 20° - 25°C, cloudy weather and relative humidity less than 80% are conducive for disease occurrence and spread.

Control measures

- Removal and destruction of diseased plant debris from the soil, proper maintenance of the crop, following crop rotation practices may help to prevent the disease occurrence.
- Spraying the crop with wettable sulphur 0.4% or mancozeb 0.2% or dinocap 0.15% or propiconazole 0.1% when initial symptoms of the disease appears, followed by 2 - 3 sprayings at 10 - 15 days intervals gives adequate control of the disease.

General characters of genus *Erysiphe* - Refer page - 159

iii) *Alternaria* blight of cumin - *Alternaria burnsii*

The disease usually occurs late during the flowering stage of the crop. Symptoms appear as scattered, dark brownish necrotic spots on the leaves and stems. The spots turn black later on when large number of colored conidia are produced on the spots. The affected stem tips bend downwards. Seed setting is adversely affected and the seeds formed are also malformed and shriveled.

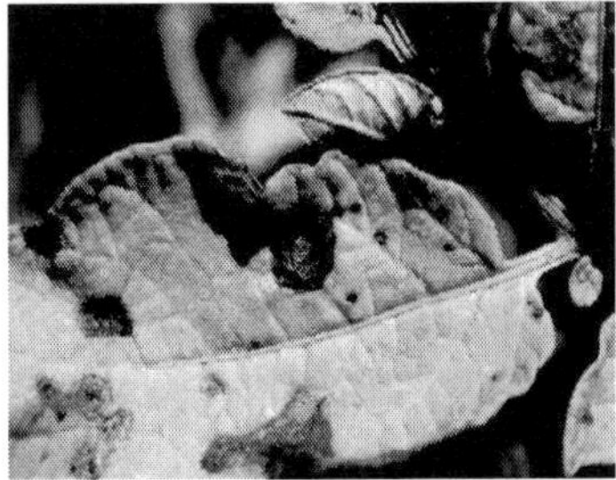
Alternaria blight of cumin

The pathogen is soil-borne, seed borne as well as air-borne. The fungus survives as dormant mycelium in the diseased plant debris in the soil and continue to produce conidia. The conidia of the fungus can also survive in the soil under warm dry conditions for several months and cause primary infection. The mycelium and spores present on the seeds also cause primary infection. The spores produced from the lesions on the standing crop are dispersed by wind, irrigation water, rain splash, farm implements etc. and cause secondary infection.

Wet weather, accompanied by high relative humidity & cloudy weather conditions & wide range of temperatures of 20°-32°C are favorable for disease development.

Control measures

- Deep summer ploughing, summer fallowing, use of disease-free seeds, collection and destruction of diseased plant debris from the soil, adoption of crop rotation practices, proper crop management etc. cam help to ward off disease occurrence.
- Spraying the crop with Bordeaux mixture 1.0% or copper oxychloride 0.25% or mancozeb 0.2% has been found to be effective in controlling the disease.

General characters of genus Alternaria

iv) Damping off of seedlings of cumin - *Pythium aphanidermatum*

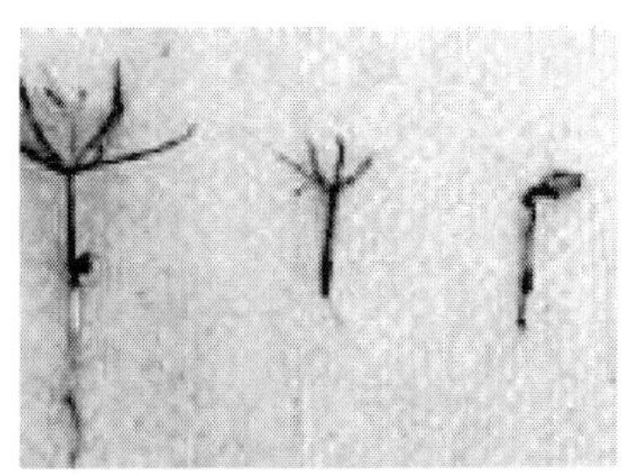

Damping off of cumin

'Damping off' occurs in two phases viz., pre-emergence and post-emergence damping off. In the pre-emergence phase, the seeds get rotted and the emerging seedlings are killed before they reach the soil surface. In the case of post-emergence damping off, soft, water soaked lesions appear near the collar region, which girdle the collar. This is followed by constriction of the collar region, and toppling of the seedlings in large numbers in patches. The young radical & plumule of the seedlings are killed and the seedlings rot completely.

The diseases is mainly soil-borne. The fungus can live in the soil as a saprophyte. The oospores produce by the fungus can also survive in the soil for several months and cause primary infection, either by producing zoospores or by germ tubes. Secondary infection may be through zoospores produced by the sporangis or by germ tubes.

High relative humidity, high soil moisture, cloudiness, and low temperatures below 24°C for a few days are conducive for the development of the disease. Crowded seedlings, dampness due to heavy rainfall, ill drained, poor and hard clay soils, application of high doses of nitrogenous fertilizers are predisposing factors for the occurrence of the disease.

Control measures

- Deep summer ploughing, locating nurseries in elevated areas with good drainage facilities, avoiding heavy clay soils for raising nurseries, raising seedlings in raised beds, avoiding thick sowing are measures that can be followed to prevent occurrence of the disease.
- Seed treatment with thiram or captan 4.0 g or metalaxyl 2.0 g / Kg of seeds provides protection to seedlings from pre-emergence damping off.
- Seed treatment with the biocontrol agents *Trichoderma harzianum* 4.0 g or *Pseudomonas fluorescens* 10.0 g/Kg of seeds protects the seedlings from infection.
- After sowing the nursery soil may be drenched with captan or thiram 0.2% or copper oxychloride 0.25% solution. This treatment may be repeated once again after emergence of the seedlings so as to cover the stem and root portions of the seedlings.

General characters of genus *Pythium* - Refer page - 133

10. Black cumin *(Nigella sativa)*

'Black cumin', 'Black seed' / 'Kalonji' is a hardy, annual, flowering plant that grows to a height of 20 - 60 cm. belongs to the Family - Ranunculaceae. The plants have fine, deeply divided leaves and the flowers are pale blue or white in color. The black seeds are borne in capsules. Black cumin is a native of Southern Europe, Northern Africa and South West Asia. It is cultivated in the Middle Eastern Mediterranean regions, Southern Europe, Northern India, Pakistan, Turkey, Syria, Iran and Saudi Arabia on a large scale. Black cumin seeds have a pungent, bitter taste and a pleasant aroma and are used as a spice in Indian and Middle Eastern cuisines. The dry, roasted seeds are used to flavor curries, vegetables and pulses. The seeds are also used as a flavoring additive in breads and pickles. Ancient Egyptians used black cumin oil as a preservative in mummification.

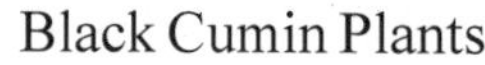

Black Cumin Plants

Black Cumin Seeds

Besides culinary uses black cumin has many medicinal uses. Traditionally the seeds are widely used for the treatment of several diseases, such as asthma, diabetes, hypertension, fever, inflammation, bronchitis, dizziness, rheumatism, skin problems, and gastrointestinal disorders. It is also used as a liver tonic, digestive and antidiarrhoeal medicine. Black cumin seeds also control parasitic infections and boost the immune systems. The medicinal properties of black cumin is mainly attributed to 'Thymoquinone', a quinine substance. The seeds contain a good amount of various vitamins and minerals, such as copper, phosphorus, zinc and iron.

Diseases of black cumin (*Nigella sativum*)

i) Wilt disease - *Fusarium oxysporum* f.sp. *cumini*

Refer wilt disease of cumin (Pages 70)

ii) Powdery mildew - *Erysiphe polygoni*

Refer powdery mildew of cumin (Pages 71)

iii) *Alternaria* blight - *(Alternaria burnsii)*

Refer *Alternaria* blight of cumin (pages 72)

iv) Damping off of seedlings - *Pythium aphanidermatum*

Refer damping off of seedlings of cumin (Pages 73)

11. Aniseed / Anise / Sweet cumin *(Pimpinella anisum)*

'Aniseed', 'Anise' or 'Sweet cumin' is an annual, herbaceous, flowering plant belonging to the Family - Apiaceae / Umbelliferae is native to the Eastern Mediterranean Region and Southwest Asia. Aniseed plants grow to a height of 3.0 meters or more. The leaves are pinnate and the plants are completely covered with fine hairs. Aniseed is widely cultivated in many countries of the world especially in Bulgaria, Cyprus, France, Germany, Mexico, Italy, South America, Syria, Turkey, Spain, UK and USSR. In India, it is grown to a small extent as a culinary herb. Aniseed is a highly prized commodity and is delightfully fragrant due to the presence of the essential oil ' anethole'.

Aniseed is extensively used for flavoring curries, bread, soups, baked products, deserts, creams, cheese, pickles, several vegetarian and non-vegetarian dishes and also in many alcoholic and non-alcoholic beverages. Aniseed has also many health promoting properties. It is rich in iron, which is vital for the production of healthy blood cells. Aniseed helps in reducing symptoms of depression. It reduces acid secretion in the stomach and helps in preventing stomach ulcer formation. It has antifungal and antibacterial properties Anethole present in aniseed helps to lower blood sugar level.

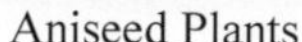

Aniseed Plants

Aniseed Seeds

Diseases of aniseed (*Pimpinella anisum*)

i) Downy mildew of Aniseed - *Peronospora umbellifarum*

Young, tender leaves of the plants are more vulnerable to attack by the disease. Symptoms of the disease appear as yellow spots or patches on the upper surface of leaves, while the under surface of the corresponding spots are covered by a white, fluffy downy growth. The lesions turn darker as they mature. In the younger leaves the whole under surface is covered by the fine, fluffy, cottony growth. The downy growth comprises of sporangiophores and sporangia, which are produced in large numbers. Later when the lesions mature, oospores may be produced and are embedded in the fluffy growth.

The fungus survives in the form of oospores released into the soil in infected plant debris. The oospores germinate by producing a germ tube and cause primary infection. Seeds contaminated with fragments of diseased plant parts may also carry the fungus. Secondary spread is through sporangia dispersed by wind or rain splash.

Low temperatures of about 14°C, high rainfall and good light are conducive for the occurrence of the disease. Prolonged leaf wetness favors disease development and spread.

Control measures

- Use of disease-free, quality seeds, removal and destruction of crop debris in the soil, removal and destruction of severely affected plants, adopting crop rotation practices, adopting field sanitation measures etc. help to minimize disease occurrence.
- Spraying with mancozeb 0.2% or metalaxyl 0.1% or a combination of mancozeb 0.2% + metalaxyl 0.1% as soon as initial symptoms of the disease appear is effective in controlling the disease.

General characters of genus *Peronospora* - Refer page - 150

ii) Powdery mildew of Aniseed - *Erysiphe heraclei*

The disease generally attacks grown up crops. Initial symptoms appear as small, irregular powdery patches on the upper surface of leaves. These powdery patches enlarge and cover large areas of the leaves, petioles, flower stalks and bracts. The affected plants appear to be covered with a powdery coating from a distance. Severely affected leaves become chlorotic and die. Severe infection causes distortion of the flowers. The white powdery coating comprises of conidiophores and conidia produced in abundance. Later, when the conidial stage is almost over, the fungus produces small, black cleistothecia, which are embedded in the mycelial mat.

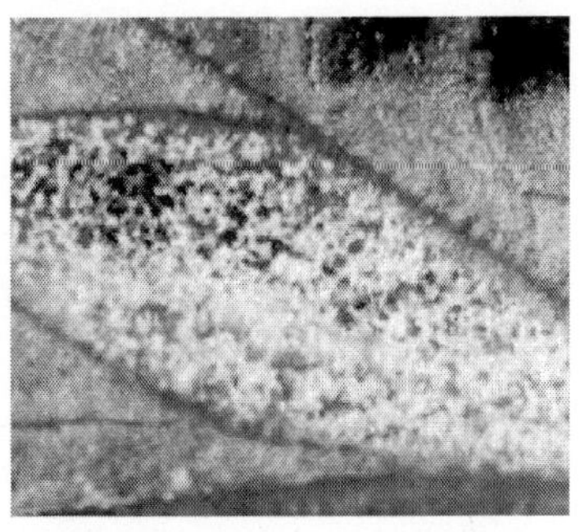
Powdery mildew of Aniseed

The disease perennates through cleistothecia in the infected plant debris in the soil. The ascospores liberated from the asci inside the cleistothecia infect the lower most leaves near the soil first. The disease is also seed-borne to some extent. Secondary spread is carried out over long distance by wind-borne conidia produced abundantly from the primary infection.

Moderate temperatures ranging from 16° - 27°C, relative humidity up to 70% and fairly dry conditions favor disease development and spread. The infection is rather high in shaded areas.

Control measures

- Removal and destruction of diseased plant debris from the soil, proper maintenance of the crop, adopting crop rotation practices etc. may help to minimize disease incidence.
- Seed treatment with captan or thiram at 4.0 g / Kg of seed eliminates seed-borne inoculum
- Spraying with wettable sulphur 0.4% or dinacap 0.15% or tridemorph 0.1% as soon as initial symptoms of the disease appears provides adequate control of the disease.

General characters of genus *Erysiphe* - Refer page - 159

iii) *Alternaria* blight of Aniseed - *Alternaria alternata*

Initial symptoms appear as small, yellow spots, scattered on the leaves. The spots enlarge and turn brown or black surrounded by a chlorotic yellow halo. Sometimes the spots appear in a concentric ringed pattern. Later the centre of the spots turns grayish-white, becomes necrotic and breaks down leaving holes. On the spots conidiophores and conidia are produced in profusion. The dark color of the conidia makes the lesions appear black. The pathogen may attack the leaf petioles and young stems and produce necrotic lesions. Severe infection causes blighting and premature defoliation of leaves.

The disease is seed-borne, soil-borne and air-borne. The mycelium and conidia found on the seeds as contaminants may cause primary infection. The fungus present in the diseased plant debris in the soil may remain active for several months and continue to produce conidia. Secondary spread occurs through air-borne conidia. Cool, moist weather conditions favor disease occurrence and spread.

Control measures

- Removal and destruction of diseased plant debris from the soil, field sanitation, proper maintenance of the crop, crop rotation etc. may help to avoid disease occurrence.
- Seed treatment with thiram or captan at 4.0 g / Kg of seed protects the crop from early infection.
- Spraying the crop with Bordeaux mixture 1.0% or copper oxychloride 0.25% or mancozeb 0.2% or captafol 0.15% or hexaconazole 0.1% or propiconazole 0.1% have been found to afford adequate control of the disease.

General characters of genus *Alternaria* - Refer page - 153

iv) Rust disease of Aniseed - *Puccinia pimpinellae*

Puccinia pimpinellae, the causal organism of rust disease of aniseed is a autoecious, micro cyclic fungus, which produces only the uredial and telial stages. Disease symptoms appear as light green, discolored lesions on the leaves, which become chlorotic. As the disease progresses, yellow-orange pustules, surrounded by a yellow halo, which are the uredia develop on the under surface of leaves. The pustules rupture and release large number of uredospores. The stems of affected plants bend and become swollen or distorted. The uredial stage is followed by the telial stage from the same pustules and teliospores are released. Telial pustules may be produced independently also. The affected plants are mostly stunted.

Rust disease of aniseed

The fungus is mainly seed-borne and the uredospores found inside the seed tissues cause primary infection. Secondary spread of the disease is carried out by wind-borne uredospores produced from the uredial pustules. High atmospheric humidity and fairly low temperatures favor disease occurrence.

Control measures

- Use of disease free seeds, good crop husbandry and crop rotation are measures that can be followed to avoid the disease.
- Seed treatment with thiram or captan at 4.0 g / Kg of seeds is effective in protecting the crop from early infection.

- Spraying the crop with wettable sulphur 0.4% or mancozeb 0.2% or chlorothalonil 0.15% or triadimefon 0.1% two or three times at 10 - 15 days interval from the time initial symptoms of thc disease is noticed controls the disease.

General characters of genus *Puccinia* - Refer page - 157

12. Mustard *(Brassica juncea)*

'Brown mustard' *(Brassica juncea)* commonly known as 'Indian mustard', 'Chinese mustard' or 'Oriental mustard' is native in the wild in Western Europe, the Mediterranean and Temperate regions of Asia and was one of the first domesticated crop. The Chinese were using it for thousands of years and the ancient Greeks considered it as an every day spice for centuries. Major producers of mustard include India, Pakistan, Canada, Nepal, Hungary, Great Britain and the United States of America. Major mustard producing states in India are Rajasthan, Gujarat, Uttar Pradesh, West Bengal, Haryana, Madhya Pradesh and Assam.

Mustard Plants

Mustard Seeds

Mustard belongs to Family-Brassicaceae (Cruciferae), Mustard is an annual, herbaceous, flowering plant with yellow flowers that grows up to a height of 4.0-6.6 feet and produces small, globular, brown colored seeds borne serially inside pods. The sulphur compounds called glucosinolates, especially Sinigrin present in mustard seeds impart the flavor and pungency to the various dishes.

Mustard seeds, leaves and flowers are all edible. Whole seeds are included in most pickling spices. Whole seeds are also toasted and used in some dishes. Powdered mustard seeds are added with other spices in the preparation of curry powders and pastes. Mustard oil is widely used in India as a culinary oil in the preparation of fish and meat dishes. Whole seeds are fried in edible oil until the seeds pop producing a mild, nutty flavor and then used as a garnish or seasoning of other Indian dishes.

Mustard has many medicinal properties also. Selenium present in mustard seeds helps in reducing severity of asthma, decreasing some of the symptoms of rheumatoid arthritis and preventing cancer. Magnesium present in mustard has been shown to help reduce the severity of asthma, to lower high blood pressure, to restore normal sleep pattern in old people, to reduce the frequency of migraine attack and to prevent heart attack in the case of diabetic heart disease. The antifungal properties of sulphur present in mustard seeds help to ward off skin infections and relieve symptoms of psoriasis. Mustard decoction has protective emetic qualities and is used to treat poisoning due to intake of narcotics or excess of alcohol.

Diseases of mustard (*Brassica juncea*)

i) *Alternaria* blight / Black spot of mustard - *Alternaria brassicae*

The disease appears on the lower leaves first and then spreads to the upper leaves. Initial symptoms appear as scattered, small, circular, brown necrotic spots on the leaf lamina. The spots gradually increase in size, coalesce and cover large areas resulting in blighting and defoliation in severe cases. Under humid conditions, a bluish, velvety fungal growth is seen on the spots. Sometimes the lesions are formed as concentric rings of brown and black. The spots become black because of the production of large numbers of colored conidia from the mycelium. Circular to linear, dark brown lesions develop on the petioles, stems and pods. Infected pods produce small, discolored and shriveled seeds.

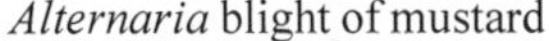

Alternaria blight of mustard

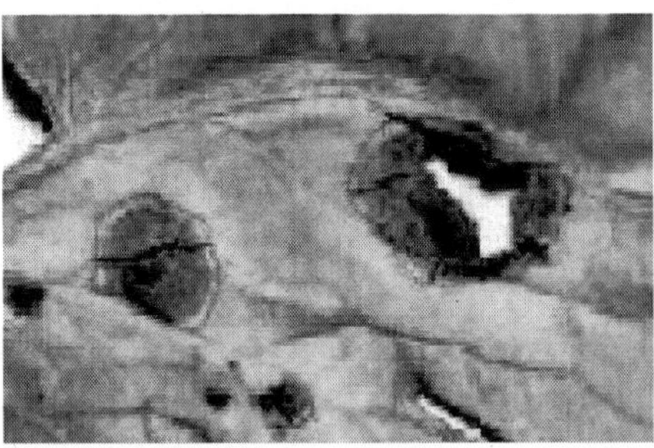

Alternaria blight of mustard

The pathogen survives as mycelium and conidia in the diseased plant debris in the soil and cause primary infection. To some extent the fungus is externally and internally seed-borne. Secondary infection takes place through wind-borne conidia produced abundantly. The pathogen has several alternate and collateral hosts and they provide enough inoculum all through the year for causing infection in mustard crop.

High relative humidity (above 70%), warm weather conditions (12° - 25°C) and intermittent rainfall favor disease occurrence and spread.

Control measures

- Removal and destruction of diseased plant debris from the soil, elimination of alternate weed hosts, crop rotation etc. may help in minimizing the chances of disease occurrence
- Seed-borne inoculum can be eliminated by seed treatment with thiram or captan at 4.0 g / kg of seeds
- Spraying with Bordeaux mixture 1.0% or copper oxychloride 0.25% or mancozeb 0.2% or captafol 0.15% provides adequate control of the disease.

General characters of genus *Alternaria* - Refer page - 153

ii) White rust of mustard - *Albugo candida*

The pathogen causes both local and systemic infection and all the parts of the infected plants exhibit symptoms. In the case of local infection shiny white or creamy yellow, isolated, raised pustules or blisters are formed on the leaves and stems in close proximity. These pustules merge to form larger lesions. Often the pustules develop in a circular formation around one or two central pustules. Usually the pustules are found on the under surface of leaves, but sometimes may appear on the upper surface also. On maturity of spores, the host epidermis is ruptured and masses of white, powdery spores are exposed.

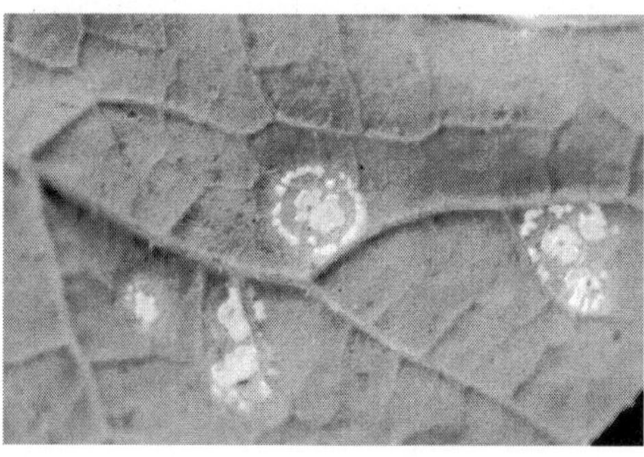

White rust of mustard

White rust of mustard

The pathogen causes both local and systemic infection and all the parts of the infected plants exhibit symptoms. In the case of local infection shiny white or creamy yellow, isolated, raised pustules or blisters are formed on the leaves and stems in close proximity. These pustules merge to form larger lesions. Often the pustules develop in a circular formation around one or two central pustules. Usually the pustules are found on the under surface of leaves, but sometimes may appear on the upper surface also. On maturity of spores, the host epidermis is ruptured and masses of white, powdery spores are exposed.

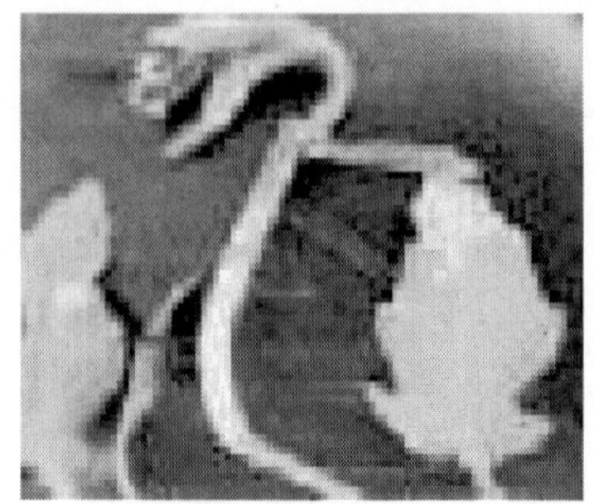

Stag head-like appearance

When young stems and inflorescence are infected, the fungus becomes systemic in the tissues and stimulates the cells, which undergo hypertrophy and hyperplasia resulting in swelling and distortion of the stems and floral parts and may present a 'stag head'-like appearance. The ovules and pollens are atrophied causing sterility. The lateral buds of infected plants may be stimulated and they grow into abnormal lateral shoots.

The pathogen produces large number of oospores in the intercellular spaces of the hypertrophied parts and perpetuates through these spores, which are found in the diseased plant debris in the soil or in diseased plant fragments along with the seeds. Primary infection is cause by motile zoospores released from the oospores. Secondary spread of the disease is through motile zoospores released from the sporangia, which are dispersed by wind and rain. The pathogen has many alternate hosts and the spores released from these hosts may attack mustard crop any time.

Moist weather conditions, high relative humidity above 70%, low temperatures below 15°C (12°-14°C) and intermittent rainfall favor disease occurrence and spread.

Control measures

- Use of disease-free, healthy seeds, removal and destruction of diseased plant debris from the soil, elimination of alternate weed hosts from the vicinity of mustard fields, eradication of severely affected plants, adoption of crop rotation practices are measures that can be followed to ward off disease occurrence.
- Seed treatment with mancozeb 4.0 g or Apron 35 SD 4.0 g / Kg of seeds is effective in controlling the disease in the seedling stage.
- Spraying with Bordeaux mixture 1.0% or copper oxychloride 0.25% or metalaxyl 0.1% or mancozeb 0.2% + metalaxyl 0.1% or fosetyl Al 0.1% has been found to afford satisfactory control of the disease.

General characters of genus *Albugo* - Refer page - 167

iii) Downy mildew of mustard - *Peronospora parasitica*

The disease usually attacks young seedlings and older leaves of grown-up plants. The symptoms appear as scattered, irregular, whitish-yellow spots or patches on the upper surface of leaves, while the under surface of the corresponding spots are covered by grayish-white, fine, cottony, downy growth. In case of severe infection the spots may be so crowded that the entire surface of the leaves are covered by the downy growth as a result the leaves shrivel, tear off easily and dry. In seedlings the entire leaf surface may be covered by the downy growth under favorable conditions for the pathogen. The stems of affected plants show deformities and small or long swellings are formed and the stalks bent abruptly. The floral buds are atrophied and seed setting is affected completely. The ovary is much elongated and turns into a twisted body. The fluffy, downy growth comprises of sporangiophores and sporangis, which are produced in abundance. Oospores, which are the sexual spores are produced in large numbers in the intercellular spaces of the tissues of the hypertrophied parts.

The pathogen survives as oospores released into the soil and in infected plant debris. Seeds contaminated with fragments of diseased plant parts may also carry the disease. Secondary spread is through sporangia dispersed by wind or rain splash. The fungus has several collateral and alternate hosts from which there may be a steady supply of inoculum, which may infect mustard crop anytime.

Low temperatures around 14°C, high rainfall of about 15 cm and good sunlight favor the development and spread of the disease.

Downy Mildew of Mustard

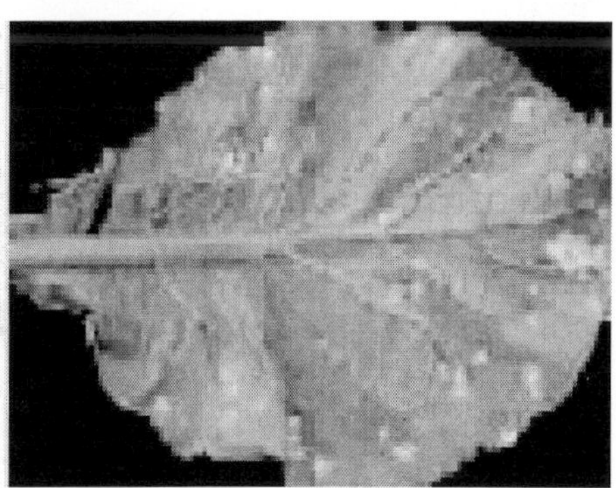

Downy Mildew of Mustard

Control measures

- Selection of disease-free, quality seeds, field sanitation, elimination of alternate weed hosts in the vicinity of mustard fields, adoption of crop rotation practices with non-cruciferous crops etc. help to minimize disease occurrence.
- Seed treatment with apron 35 SD 4.0 g or thiram 4.0 g or captan 4.0 g / kg of seeds is effective in protecting the seedlings from infection.

- Spraying with metalaxyl 0.1% or captafol 0.15% or ziram 0.2% has been found to afford good control of the disease.

General characters of genus *Peronospora* - Refer page - 150

iv) Powdery mildew of mustard - *Erysiphe cruciferarum*

The disease usually attacks the older leaves to a larger extent than younger leaves. Plants grown under stressed conditions are more susceptible to attack by the disease. The symptoms appear as small, discrete, whitish patches on both the surfaces of leaves. These patches coalesce and cover large areas of the leaf surface. The patches are covered by a mass of white, powdery fungal growth comprising of mycelium, conidiophores and conidia. The powdery patches may cover the petioles, stems and even the pods. The affected plants appear grayish-white from a distance. Eventually the leaves turn yellow, then brown, become necrotic and drop off prematurely. Later in the season the powdery mildew growth is interspersed with numerous, small, black cleistothecia, which are the sexual fruiting bodies of the fungus.

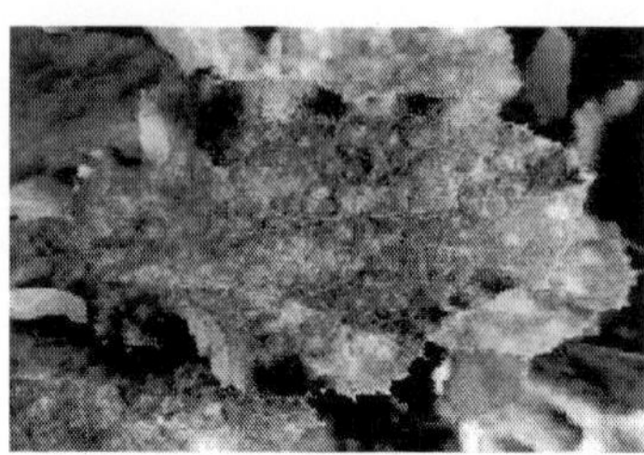

Powdery mildew of mustard

The disease perennates through cleistothecia in the diseased plant debris in the soil. The ascospores liberated from the asci inside the cleistothecia initiate primary infection on the lower leaves near the soil first. Fragments of diseased plant materials adhering to the seeds may also carry the pathogen and cause primary infection. Secondary spread is carried out by means of wind-borne conidia produced in abundance from the aerial mycelium. The pathogen has several alternate cultivated and weed hosts and so there is a steady supply of spores that can infect mustard crop any time.

Cool nights and warm days with temperatures ranging from 16° - 27°C, relative humidity up to 70% and low rain fall favor disease development and spread.

Control measures

- Removal and destruction of diseased plant debris from the soil, elimination of alternate weed hosts from the vicinity of mustard fields, provision of drought stress free conditions, adoption of crop rotation practices, field sanitation etc. minimize chances of disease.
- Treating the seeds with captan or thiram 4.0 g / Kg of seeds is effective in eliminating seed-borne inoculum.

- Spraying with wettable sulphur 0.4% or dinocap 0.15% or triadimefon 0.1% or hexaconazole 0.1% has been found to be effective in controlling the disease.

General characters of genus Erysiphe - Refer page - 159

v) Club root of mustard - *Plasmodiophora brassicae*

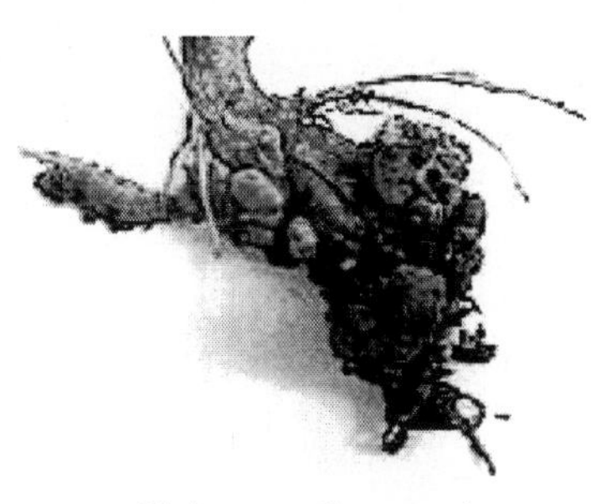
Club root of mustard

'Club root' or' finger and toe disease' attacks several cultivated and weeds belonging to Family - Cruciferae. Club root is primarily a root disease and no visible symptoms of the disease are manifested on the aerial parts of the infected plant in the early stages of infection. The first above ground symptoms are yellowing and wilting of leaves during the day time as seen under water stress conditions, as well as unthrifty, sickly appearance of the plants. When such plants are uprooted, galls and club-like swellings are found on the tap root and lateral roots. Numerous resting spores are formed inside the swollen parts, which are released into the soil after disintegration of the galls.

The club root pathogen is primarily soil-borne. The resting spores can remain in a viable state for many years in the infected plant debris in the soil. The spores may also becarried on the seed. The spores may be spread through soil, rain and irrigation water, strong wind, farm implements and hooves of cattle. The dung of cattle fed with raw club roots also contain the spores in a viable state. The resting spores germinate when a suitable host is present and release zoospores. From the zoospore a plasmodium is released. The plasmodium enters a young host root by direct penetration or through some injuries caused in the root. The presence and activity of the plasmodium inside the host cell leads to proliferation of the occupied cell as well as the adjacent cells leading to formation of galls and clubs on the roots. The galls and clubs restrict uptake of water from the soil by the host plants and thereby causes wilting.

Soil moisture levels above 60% and temperatures ranging from 27° - 30°C favor disease development and spread. The incidence of the disease is more in acidic soils.

Soil moisture levels above 60% and temperatures ranging from 27° - 30°C favor disease development and spread. The incidence of the disease is more in acidic soils.

Control measures

Removal and destruction of diseased roots and plant debris from the soil, provision of proper drainage facilities to avoid water logging, elimination of alternate weed hosts in the vicinity of mustard fields, avoiding raising of mustard in infected sick soils, use of seedlings raised from disease-free soils for planting, avoiding feeding of raw club roots to cattle, application of hydrated lime or gypsum so as to raise the soil pH to about 7.0, adoption of long term crop rotation practices with non-leguminous crops, preventing movement of cattle from infected to other fields etc. are measures that can be followed to minimize disease occurrence and spread

Treatment of seed beds with soil fumigants, such as methyl bromide - 1.0 lit. or chloropicrin - 1.0 lit. or vapam - 1.3 lit. / cent of nursery area with a soil injector at a depth of 15 - 20 cm. and at 30 cm interval, 2 - 3 weeks before sowing affords good control of the disease. Drenching the seed bed with formalin 2.0% at the rate of 10 lit. / cent of nursery area, 3 - 4 weeks before sowing is also effective in controlling the disease.

General characters of genus *Plasmodiophora* - **Refer page - 169**

(vi) *Sclerotinia* stem rot / White mould of mustard - *Sclerotinia sclerotiorum*

The disease attacks the crop mostly during the flowering and maturity stages. Symptoms appear as elongated, soft, water-soaked, white to gray lesions on the leaves and stems near the crown region. The lesions enlarge and girdle the stem. Plants above the affected area turn pale green or yellow, wilt and eventually die. Mature lesions become bleached at the internodes. Affected plants mature quite early. The stems shred easily resulting in lodging. White mould grows on the rotting stems and brown to black sclerotia are produced inside the affected stems.

The pathogen survives as mycelium in the dead or live plants and as sclerotia in the infected plant parts in the debris in the soil or in contaminated seeds. Under favorable conditions apothecia are formed at the end of long stalks from the sclerotia and ascospores are released from asci found inside the apothecia. The ascospores, which are carried through wind, water, soil and other agents cause fresh infection.

High relative humidity ranging from 90 - 95% and moderate temperatures of 18° - 25°C and blowing wind are favorable for disease development and spread.

Control measures

Removal and destruction of diseased plant debris from the soil, deep summer ploughing, use of disease-free seeds, removal and destruction of alternate weed hosts from the vicinity of mustard fields, adoption of long term crop rotation, removal of severely affected plants etc. help to minimize disease occurrence and spread.

Seed treatment with captan or thiram at 4.0 g or carbendazim at 2.0 g / kg of seeds is effective in warding off the disease in young plants.

Spraying with carbendzim 0.1% or topsin M 0.2%, 50 - 60 days after sowing has been found to be effective in controlling the disease to some extent.

General characters of genus *Sclerotinia* - Refer page - 161

(vii) Bacterial blight / Black rot of mustard - *Xanthomonas campestris* pv. *campestris*

Symptoms of the disease appear as irregularly shaped, dull yellow, chlorotic areas along the leaf margins, which expand towards the leaf mid rib and form a characteristic 'V'- shaped lesion with the base of the 'V' towards the mid rib. Lesions may coalesce along the leaf margins and the plants present a scorched appearance. The veins show brown to black discoloration. Dark colored streaks are formed on the stem from the ground level and the streaks enlarge and girdle the stem. The affected stem becomes hollow due to internal rotting. Midrib of lower leaves may crack and the veins turn brown and the leaves wither. In severe cases the vascular bundles of the stem turn brown and the plant collapses.

The bacteria can survive in the infected plant debris in the soil as a saprophyte and are also present on the seeds. Infected seeds may cause primary infection. The bacteria may spread through irrigation water, rain splash and by strong winds along with diseased plant materials The bacteria present in the diseased plant debris gain access to the plants through hydathodes mostly. Infection may take place through other natural openings on the leaves as well as through injuries.

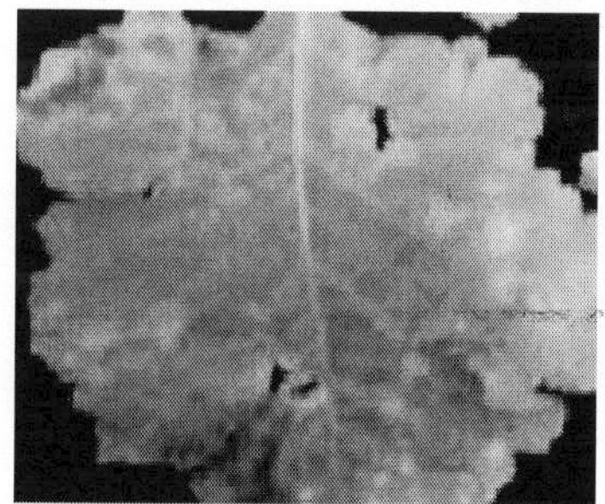

Bacterial blight of mustard

Warm and humid weather conditions favor disease development and spread.

Control measures

Removal and destruction of diseased plant debris from the soil, use of disease-free seeds, proper crop management, adoption of crop rotation practices etc. help to prevent occurrence of the disease.

Seed treatment with Auriomycin 100 - 0.1% solution for 30 minutes eliminates both internally and externally seed-borne bacteria.

Hot water treatment of seeds at 50°C for a continuous period of 30 minutes eliminates the bacteria present in the seeds.

Drenching the seed beds with Agrimycin 100 - 0.6 g / litre of water at the rate of 10 litres / cent nursery area prior to sowing controls the disease effectively.

Spraying the crop with Agrimycin 100 - 0.6 g / litre of water gives good control of the disease.

Spraying the crop with copper oxychloride 0.25% prevents secondary spread of the disease.

13. Black mustard *(Brassica nigra)*

'*Brassica nigra*' commonly known as **'black mustard'** belonging to Family - Brassicaceae (Cruciferae) is an annual flowering herbaceous plant. It is probably native to Northern Africa, Western and Central Asia and parts of Europe. It grows to a height of 40 - 80 cm and produces yellow flowers. The seeds are borne in long seed pods that contain four, dark brown or black, rounded seeds. It is widely cultivated all over Europe and the United States of America.

Black mustard plants

Black mustard is a spice that imparts a specific flavor to the foods and is commonly used in Indian cuisines, such as curries and other non-vegetarian dishes. The seeds are usually thrown into hot oil or ghee after which they pop releasing a characteristic nutty flavor and is used to garnish various food items. Mustard is mainly used to flavor several recipes of fish, meat, sauces, salads and snacks. The seeds have a good amount of fatty oils mainly oleic acid. The oil is used often as cooking oil in India.

Black mustard seeds

Black mustard has many medicinal properties. Black mustard oil is used for the treatment of common cold, painful joints and muscles (Rheumatism) and arthritis. Ground black mustard seed is used for inducing vomiting, relieving water retention (Edema) by increasing urine secretion and increasing appetite. Warm black mustard seed paste packed in cloth is applied on the skin as a mustard plaster to treat pain and swelling (Inflammation) of the lining of the lungs (Pleurisy), pneumonia, arthritis, lower back pain (Lumbago) and aching feet.

Diseases of black mustard (*Brassica nigra*)

Most of the diseases that attack brown mustard / Oriental mustard attack black mustard also.

i) Black spot/*Alternaria* blight *(Alternaria brassicae)*
   Refer *Alternaria* blight of brown mustard (Pages 80)

ii) White rust *(Albugo candida)*
   Refer white rust of brown mustard (Pages 81)

iii) Downy mildew *(Peronospora parasitica)*
   Refer downy mildew of brown mustard (Pages 82)

iv) Club root *(Plasmodiophora brassicae)*
   Refer club root of brown mustard (Pages 85)

v) *Sclerotinia* stem rot / White mould *(Sclerotinia sclerotiorum)*
   Refer white mould of brown mustard (Pages 86)

vi) Powdery mildew *(Erysiphe cruciferarum)*
   Refer powdery mildew of brown mustard (Pages 84)

vii) Black rot / Bacterial blight *(Xanthomonas campestris* p.v. *campestris*
   Refer bacterial blight of brown mustard (Page 87)

14. White mustard / Yellow mustard *(Sinapis alba / Brassica alba)*

'White mustard', which belongs to Family - Brassicaceae is virtually a wild plant and has probably originated in the Mediterranean regions. Now it is wide spread all over the world and is grown in almost all the countries especially Europe, North Africa, the Middle East and Central Asia for its seed and other vegetative parts. It is also grown as a cover and fodder crop.

White mustard plants

White mustard is a hardy, annual, herbaceous plant that grows up to a height of about 70 cm. The plants produce yellow colored flowers with four petals and the pods are bristly. Each pod bears about half a dozen hard, light yellow, globular seeds.

The seeds are used whole for pickling or toasted for use in various dishes. The seeds add a pleasant flavor and a pungent taste. However, the flavor and pungent taste is milder compared to that of the other mustards. The seeds are used to make a condiment called 'mustard'. The seeds, which contain a thioglycocide called 'Sinalbin' is responsible for imparting the pungent taste and flavor.

White mustard seeds

Besides culinary uses, white mustard has many medicinal properties. The seeds possess antifungal, antibacterial, emetic, diuretic, digestive, expectorant, stimulant, appetizing, carminative and rubifacient properties. White mustard seeds are also used for treating arthritic joints, skin eruptions and respiratory infections. In China white mustard leaves are used to treat coughs, pleurisy as well as tuberculosis.

Diseases of white mustard (*Sinapis alba*)

White mustard plants are quite hardy compared to brown and black mustards and so they are not vulnerable to the diseases which attack other mustards commonly. However, under highly favorable conditions for the pathogens the diseases viz., Black spot *(Alternaria brassicae),* Downy mildew *(Peronospora parasitica),* White rust *(Albugo candida),* White mould *(Sclerotinia sclerotiorum),* Club root *(Plasmodiophora brassicae),* Powdery mildew *(Erysiphe cruciferarum)* and Black rot *(Xanthomonas campestris* pv. *campestris)* that attack other mustards may attack white mustard also.

Refer diseases of brown mustard - Page 80-89.

15. Cinnamon *(Cinnamomum verum /C. zeylandricum)*

The genus *Cinnamomum* belonging to family - Lauraceae comprises of over 300 species distributed in the tropical and subtropical regions of many countries. Some of these species are grown commercially as spices. *Cinnamomum verum* popularly known simply as 'cinnamon' or 'Ceylon cinnamon', a native of Sri Lanka and Southern India was subsequently grown in Java, Sumatra, Borneo, Mauritius and Guyana. Besides these countries, now it is grown in South

Cinnamon tree

America, West Indies and in other tropical and subtropical countries commercially. The leaves and bark of these plants have aromatic oils.

Cinnamon quills

Cinnamon, a small, bushy, evergreen tree grows to a height of 10 - 15 meters and its bark is widely used as a spice all over the world. The flowers of this plant, which are produced in panicles have a greenish color. The fruit is a purple colored berry with a single seed. The flavor of this spice is due to an aromatic essential oil of cinnamic aldehyde or cinnamaldehyde, which gives off the characteristic odor of cinnamon and very hot aromatic taste. The essential oil contains ethyl cinnamate, eugenol, cinnamaldehyde and other chemical compounds.

Cinnamon is harvested after growing the tree for two years and then coppicing it. From the cut stem about a dozen shoots will emerge, which will provide cinnamon for the succeeding harvest after about two years. The cut shoots are stripped off their bark and the inner thin bark is peeled off. Only this thin inner bark is used as a spice. The inner bark on drying under the sun naturally curls into quills, which are cut into desired sizes for marketing.

Diseases of cinnamon (*Cinnamomum verum*)

i) Leaf spot and die back of cinnamon - *Colletotrichum gloeosporioides*

Leaf spot of cinnamon

'Leaf spot and die back' is a serious disease of cinnamon and is prevalent in almost all the cinnamon growing countries of the world. The disease affects the crop at all growing stages. When seedlings are affected they dry and die prematurely. Initial symptoms appear on leaves of grown up plants as small, brown specks, which later coalesce to form large, brown necrotic patches and the leaves present a scorched appearance. Sometimes the spots become papery with a dark brown margin at a later stage, The central necrotic areas of the spots break down and fall off leaving shot hole symptoms. Infection may extend to the stem region causing die back of the affected stem. Severe infection may cause serious yield loss.

Etiology, Epidemiology and control measures - Refer diseases of pepper (Page 15)

ii) Canker / Stripe canker of cinnamon - *Phytophthora cinnamomi*

'Stripe canker' is a disease that attacks the shoots and young stems of cinnamon. Vertical stripes in the form of sunken cankers are formed on the stem and amber colored exudates ooze out at the advancing margins of the cankers and harden. Vertical stripes of dead bark are more numerous near the ground level.

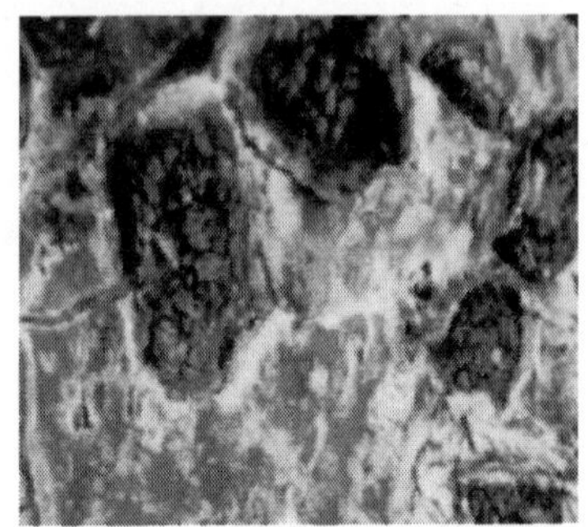

Stripe canker of cinnamon

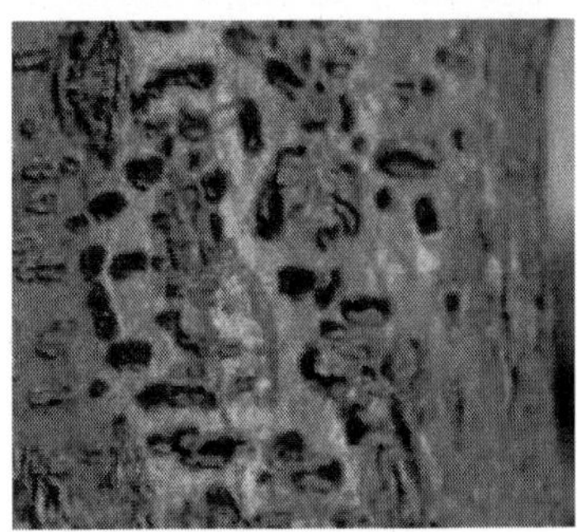

Stripe canker of cinnamon

The causal organism is a facultative parasite. It can survive in the soil as asexually formed, thick-walled resting spores called chlamydospores or sexually formed thick-walled resting spores called oospores or as vegetative mycelium in infected plant debris. The disease spreads through air-borne sporangia produced from mycelium in the host as well as through water splash.

The disease is prevalent in ill-drained soils. Low temperatures between 20° - 28°C and high relative humidity near about 100 per cent is favorable for disease occurrence and spread. The motile zoospores require water droplets on the host surface for germination and successful penetration and entry into the host.

Control measures

Improving soil drainage conditions, use of disease-free seeds and planting materials, avoiding planting in water saturated soils, adopting phytosanitation methods, such as removal and destruction of affected parts are methods that can be followed to avoid disease incidence and spread.

Wound dressing with tar or Bordeaux paste, soil drenching with Bordeaux mixture 1.0% or copper oxychloride 0.25% are methods that can be followed to minimize the disease severity.

General characters of genus *Phytophthora* - Refer page - 142

iii) Grey leaf spot / Blight of cinnamon - *Pestalotia cinnamomi*

'Gray leaf spot' or blight of cinnamon is one of the common diseases attacking cinnamon *Pestalotia palmarum* is also associated with the disease. Initial symptoms of the disease appear as small, scattered, yellowish-brown spots on the leaves. Later these spots enlarge and turn gray with a dark border and spread over the entire leaf lamina and may cause severe damage and defoliation. Sometimes the pathogen infects the twigs near the petiole. Dark acervuli are produced on the center of older lesions, which appear as black, pin point-like dots.

Grey Blight of cinnamon

The conidia are dispersed mostly by wind. When weather conditions are unfavorable for the pathogen and when suitable hosts are not available, the pathogen remains dormant in the infected plant parts or in the infected plant debris in the soil.

High atmospheric moisture and moderate temperatures favor the occurrence and spread of the disease. Plants weakened by other causes, such as insect attack, inadequate water supply, over dose of chemical fertilizers, water-logging etc. predispose the plants to attack by the pathogen.

Control measures

Field sanitation, collection and destruction of diseased plant debris periodically, application of recommended dose of fertilizers etc. may help to minimize the disease occurrence and spread.

Spraying with Bordeaux mixture 1.0% or copper oxychloride 0.25% or dithiocarbamates 0.2% is effective in controlling the disease.

General characters of genus *Pestalotia* - Refer page - 171

iv) Pink disease of cinnamon - *Corticium salmonicolor / Erythricium salmonicolor*

The symptoms of the disease appear as thin, pale pinkish-white encrusted areas on stems or branches in the beginning. Later the infection spreads destroying the bark and finally leading to the death of slender shoots. The main visible signs of the disease is the sudden death of a branch, with the brown, dry leaves hanging down but still attached to the branches by means of fungal chords. The basidiocarps appear as a very thin encrustation resembling a coating of drab paint on a twig or branch.

The disease spreads mostly through air-borne basidiospores. The fungus also spreads through mycelium present in diseased materials, which are carried by wind currents. The spores produced in alternate hosts also spread the disease in cinnamon.

The disease persists in the infected plants for years and causes gradual deterioration of the affected plants. In badly ventilated fields, where air movements are restricted by wind barriers, under dense shade or areas surrounded by thick vegetation, such as jungles, where the air in the rainy seasons is hot, moist and still the disease incidence is more. Pink disease is considerably more in badly maintained gardens.

Control measures

Proper maintenance of the plants by the application of recommended dose of fertilizers, pruning and destruction of severely affected branches and adoption of other plant sanitary measures help in mitigating the disease.

Application of copper-based fungicidal paste over the cut ends of pruned branches and spraying with Bordeaux mixture 1.0% or copper oxychloride fungicides 0.25% when initial symptoms of the disease appear have been recommended for controlling the disease.

General characters of genus *Corticium* - Refer page - 172

16. Cassia *(Cinnamomum cassia / C. aromaticum)*

'Cassia' *(Cinnamomum aromaticum)*, an ever green tree is native to Southern China and South East Asia. Its flavor is less delicate than that of true cinnamon or Ceylon cinnamon. Whole branches and small trees are harvested for cassia bark, unlike the small shoots used in the production of true cinnamon. Cassia cinnamon is rather rough and thick unlike the thin true cinnamon. Cassia bark either powdered or as sticks is used as a flavoring agent for candies, desserts, baked goods, meat preparations and in many curry recipes. Cassia buds are also used as a spice in India.

Cassia tree

Cassia quills

Diseases of cassia (*Cinnamomum cassia*)

All the diseases that attack cinnamon attack cassia also.

i) Leafspot and die-back - *Colletotrichum gloeosporioides*
   Refer leaf spot and die-back of cinnamon (Page 91)

ii) Canker / Stripe canker - *Phytophthora cinnamomi*
   Refer canker / stripe canker of cinnamon (Page 92)

iii) Grey leaf spot / Blight - *Pestalotia cinnamomi*
   Refer grey leaf spot / blight of cinnamon (Page 92)

iv) Pink disease - *Corticium (Erythricium) salmonicolor*
   Refer pink disease of cinnamon (Page 93)

17. Allspice *(Pimenta dioica)*

Allspice belonging to Family - Myrtaceae of the Plant Kingdom is an evergreen shrub or a small scrubby tree that grows up to a height of 10 - 18 meters. Allspice plant is native to the Greater Antilles, Southern Mexico and Central America. Now it is widely cultivated in almost all countries of the world, especially in the warm regions. Jamaica is the largest producer and exporter of allspice accounting for 70 per cent of world trade. Allspice is also known as 'Jamaica pepper', 'Myrtle pepper', 'Pimenta' and 'Pimento'. It is a spice that combines the flavor of cinnamon, nutmeg and clove.

Allspice tree

Allspice is actually the dried, unripe berry of *Pimenta dioica.* The fruits are picked when green and unripe and are dried in the sun. The dried berries are brown in color and resemble large, smooth pepper corns. The leaves are also used in cooking. Further, the leaves and wood are often used for smoking meat. Allspice is used as an ingredient in commercial sausage preparations and curry powders. In the Middle Eastern countries allspice is extensively used to flavor a variety of stews and meat dishes. In Arab cuisines it is the sole spice added for flavoring. In the West Indies, 'Pimento dram' an allspice liquor is produced. In the United States it is used mostly in desserts. In Great Britain it is used in various dishes including cakes. It is also used in the manufacture of many beauty products. In Portugal whole allspice is widely used in many traditional dishes and stews. In India it is used in some vegetarian and chicken curries. It is one of the main ingredient in pickles in many countries.

Besides culinary uses, allspice is used as medicine to treat many ailments. It is found to have anti-inflammatory, rubifacient, carminative and anti-flatulent properties. It aids in digestion by increasing enzyme secretion in the stomach and intestines. 'Eugenol' present in allspice has local anesthetic and antiseptic properties and is found useful in gum and dental treatment. The spice contains good amount of minerals, such as potassium, manganese, iron, copper, selenium and magnesium. It also contains a good amount of Vitamin A, Vitamin B-6, Riboflavin, Niacin and Vitamin C, which are quite vital for the physiological functioning of human body.

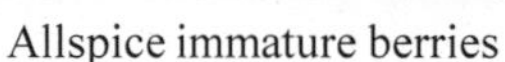

Allspice immature berries

Mature berries

Diseases of allspice (*Pimenta dioica*)

(i) Leaf rust / Myrtle rust / Guava rust of allspice - *Puccinia psidii*

Leaf rust' of allspice commonly known as 'Myrtle rust' or "Guava rust' is a serious disease and causes heavy damage to the plants and even death of the affected plants. It has a wide host range and attacks many species belonging to the family Myrtaceae.

Myrtle rust of allspice

The first sign of infection is the appearance of small chlorotic specks or lesions. These lesions mainly appear as pustules on young, actively growing leaves and shoots. Such lesions also occur on flowers and fruits. This is followed by the production of masses of bright yellow urediniospores, which cover the affected parts as a powdery coating and present a rusty appearance. Lesions are formed on both the upper and lower surfaces of leaves. Lesions often turn reddish-purple, then grey with age and have a purple or dark brown margin. They tend to become angular, coalesce and extend over the entire leaf area. Severe infection results in defoliation, leading to the death of the tree within 3 - 4 years.

The disease spreads only through urediniospores. The spores are mostly spread by wind even to very distant places. Honey bees, mammals, such as squirrels and some birds are also involved in the dispersal of spores to some extent.

Disease development is favored by low temperatures, around 20°C, high relative humidity, around 80% during night time and leaf wetness. A period of darkness is necessary for successful infection. Spore germination is more at temperatures between 13° - 22°C and leaf wetness.

Control measures

Quarantine laws have been enacted to prevent movement of plants and plant parts from disease affected countries to other countries and also from disease affected parts to other parts where the disease is not prevalent.

Judicious pruning and cutting and destroying severely infected branches are methods that can be followed to restrict the spread of the disease

Spray application with Bordeaux mixture 1.0% or Copper oxychloride 0.25% or Mancozeb 0.2% or Chlorothalonil 0.15% or Propiconazole 0.1% is effective in controlling the disease. Spraying should commence as soon as disease symptoms are noticed and repeated sprayings are required at fortnightly intervals.

General characters of genus *Puccinia* - Refer page - 157

ii) Die-back / Canker of allspice - *Ceratocystis fimbriata*

'Die-back / canker' is another important disease of allspice, which may cause even death of affected trees. The disease is usually localized and gradually spreads to all other parts of the tree. Infected tree displays bark canker, dark streaks in the wood and drying of leaves. Severely affected trees die within a few months when primary infection occurs low in the tree trunk compared to infection through the branches.

The fungus survives as mycelium within the host plant and in the cankers or as ascospores in the host or in the soil or debris. The fungus may be dispersed as fragments of mycelium along with dead bark pieces and as chlamydospores or conidia or ascospores produced in perithecia. The hat-shaped ascospores are thick-walled and survive for long periods in soil and in insect frass. Dispersal of ascospores is mostly done by insects. Conidia found in insect frass may be dispersed by wind.

Temperatures ranging from 18° - 28°C is optimum for growth and ascospore production by the fungus

Control measures

Once the disease is established and cankers are formed, it is very difficult or rather impossible to control the disease. Collection and destruction of fallen dead branches of trees, barks of trees and debris from the soil, judicious pruning and cutting of infected branches help in preventing further spread of the disease. After pruning the cut ends should be treated with a copper based fungicidal paste. The disease may spread through pruning knives and other cutting tools also and so the tools should be cleaned with disinfectants frequently. As the disease is a typical wound parasite, every effort should be taken to prevent causing wounds either by insects or by human activities.

General characters of genus *Ceratocystis* - Refer page - 173

Diseases of nutmeg (*Myristica fragrans*)

'Nutmeg' is a tropical, evergreen tree belonging to Family - Myristicaceae. It is a native to Indonesia and now it is grown in most of the tropical countries of the world. Guatemala and Indonesia combined produce about 68 per cent of the world supply of nutmeg. Nutmeg is the spice made by grinding the seeds of the fragrant nutmeg tree into a powder. The spice has a distinctive pungent fragrance and a warm, slightly sweet taste. It is used to flavor many kinds of baked goods, confectionaries, puddings, potatoes, meats, sausages, sauces, vegetables and many kinds of beverages. It is also used in juices, pickles and chutnies.

The essential oil extracted from nutmeg contains terpenes and phenylpropanoids and is used in perfumery and pharmaceutical industries, especially in the manufacture of tooth pastes and cough syrups. The oil is also used as a natural food flavoring agent in baked goods, syrups, beverages and sweets.

Nutmeg tree

Nutmeg fruit

Besides culinary uses, nutmeg has many medicinal uses. Nutmeg is used to treat insomnia by inducing sleep. It helps in digestion and relieves stomach problems. It improves blood pressure and restores normal blood circulation. It stops tooth aches and prevents bad breath. The volatile oils in nutmeg, such as myristicin, elemicin, eugenol and safrole relieve inflammation, swelling, joint pain, muscle pain and sores.

Diseases of nutmeg

i) Leaf spot and shot hole of nutmeg - *Colletotrichum gloeosporioides*

Leaf spot and shot hole of nutmeg

Initial symptoms appear as small, sunken spots, surrounded by a yellow halo on the leaves. The spots enlarge to some extent and subsequently the central portion of the necrotic region drop off resulting in shot hole symptom. Die-back symptoms are also evident in some of the mature branches. On young seedlings drying of the leaves and subsequent defoliation are seen.

Etiology, epidemiology and control measures and general characters of genus *Colletotrichum* - Refer anthracnose of pepper (page 15)

ii) Fruit rot of nutmeg - *Colletotrichum gloeosporioides* and *Diplodia (Botryodiplodia) theobromae*

Fruit rot of nutmeg

The symptoms appear as water soaked lesions on the fruits. On immature fruits splitting of the pericarp and rotting are seen. Affected fruits drop off prematurely. In case of fruit rot, the infection starts from the pedicel as dark lesions and gradually spreads to the fruits causing brown discoloration of the rind. This is followed by rotting of the mace and the seeds, and the tissues become discolored and disintegrated. The rotten tissues emit a foul odor. The fallen fruits become enveloped with the growth of the pathogens.

Etiology, epidemiology and control measures and general characters of genus *Diplodia* - Refer fruit rot of nutmeg

19. Vanilla *(Vanilla planifolia / V. tahitensis)*

The vanilla orchids form a flowering plant genus comprising of about 110 species in the orchid family - Orchidaceae. The most widely known member is the flat-leafed vanilla - *Vanilla planifolia*, native to Mexico from which commercial vanilla flavoring agent is derived. Vanilla is widely used for industrial purposes in flavoring products, such as foods, beverages, particularly chocolates, confectionary items, ice creams and bakery goods, and in cosmetics, perfumery etc. The most popular vanilla aroma and flavor is imparted by the phenol aldehyde 'Vanillin'. Vanilla is also obtained from Tahiti vanilla - *Vanilla tahitensis* and West Indian vanilla - *Vanilla pompona*

Vanilla plants

Vanilla plant with pods

Dry vanilla pods

Vanilla seeds

The ever green genus occurs world wide in the tropical and sub tropical regions of America, Asia, New Guinea and West Africa. The vine-like vanilla are climbing plants. They have long, thin stems with alternate leaves. The short, oblong leaves are flat, thick and leathery. Long and strong aerial roots grow from each node. Flowers arise on short peduncles from the leaf axis. The flowers are quite large, attractive and are whitish, greenish or greenish-yellow or cream colored and are sweet smelling. Vanilla fruit is an elongate, dehiscent capsule. After flowering, it takes 8 - 9 months for the capsules to ripen. Each capsule contains thousands of minute seeds and both the pods and seeds are used to obtain the vanilla flavoring agent.

Apart from being a flavoring agent, vanilla has several therapeutic properties. Vanilla acts as an antioxidant, anticarcinogenic, antidepressant and aphrodisiac, and provides many health benefits.

Diseases of vanilla (*Vanilla planifolia*)

i) Stem rot of vanilla - *Fusarium oxysporum* f.sp. *vanillae*

'Stem rot' is a very serious disease of vanilla and is responsible for causing heavy yield loss The symptoms appear as water-soaked lesions on the stem, which later enlarge and girdle the stem. The affected region turns brown in color. Subsequently elongated, water-soaked patches develop on the stem. The tissues from such from water-soaked lesions and the tissues in that region rot. In advanced stages, the leaves turn yellowish and dry. When the stem in the basal or middle portions of the vines are infected, the affected regions shrivel and the distal portions above show wilting symptoms even when enough water is available. The stem infection is often followed by rotting of the capsules.

Stem rot of vanilla

The pathogen is mostly soil-borne. The mycelium can survive in the diseased plant parts in the soil for a long time and continue to produce conidia and the conidia dispersed by wind and rain splash may cause fresh infection. The chlamydospores produced in large numbers can survive for long periods of time, even for months together in the soil and in the infected plant debris. They are capable of withstanding unfavorable climatic conditions, remain viable and can cause fresh infection.

Excessive soil moisture, too much shading to the vines, inadequate fertilization, over crowding and drought stress predispose the vines to infection and spread of the disease.

Control measures

Collection and destruction of plant debris from the soil, use of disease-free planting materials, avoiding over crowding of vines, proper maintenance of the vines by judicious application of fertilizers and timely irrigation, avoiding water stagnation by providing suitable drainage facilities, avoiding excessive shading to the vines, cutting and destroying severely affected vines etc. are measures that can be followed to minimize disease occurrence and spread.

Soil drenching with carbendazim 0.2% or mancozeb 0.2% or a combination of the two fungicides or copper oxychloride 0.25% or Bordeaux mixture 1.0% affords good control of the disease.

Application of biocontrol agents, such as *Trichoderma viride* or *Trichoderma harzianum* or *Pseudomonas fluoroscens* has been found to inhibit the pathogen.

General characters of genus *Fusarium* - Refer page - 138

ii) Root rot of vanilla - *Fusarium batatis* var. *vanillae*

'Root rot' of vanilla is another serious disease of vanilla that causes severe root damage. The infection starts as browning of the underground root tips, which soon decay and result in the death of underground roots. The infection gradually spreads to the aerial roots also. The infected plants produce numerous aerial roots, but most of these roots die before they reach the soil. The stem and leaves of the affected vine become flaccid, the stem shrivels and the affected vine presents a drooping appearance even when sufficient water is available. The fungus, as well as the conidia produced in large numbers block the water conducting vessels leading to wilting of the vines.

Etiology, epidemiology and control measures - Refer Stem rot of vanilla (Page 101)

iii) Shoot tip rot of vanilla - *Colletotrichum gloeosporioides*

The symptom of the disease is rotting of shoot tips of the main vine or the side branches. Initial symptoms appear as rotting in the form of brown patches on the petiole and lower portion of the youngest leaf. The unfurled youngest leaf, which is in the form of a funnel retains rain water or water from dew drops and facilitates infection. Within a few days the rotting extends to the affected leaf and the shoot tip as a result the shoot tip becomes soft, changes into brown color, die and drop off. The infection is confined to the shoot tips only.

Shoot tip rot of vanilla

Diseased, fallen shoots on the ground continue to produce conidia and the wind-borne conidia serve as a source of fresh infection. The fungus can also survive saprophytically on the plant debris in the soil. Conidia produced from rotten shoots, which have not fallen off can also cause fresh infection. Further, the pathogen has many other alternate hosts. The spores produced from such hosts may also cause infection on vanilla.

The disease occurs mostly during the post monsoon periods (September December). High relative humidity of 95 - 97%, misty conditions and moderate temperatures of about 25°C are favorable for the disease occurrence and spread.

iv) Inflorescence rot of vanilla - *Colletotrichum vanillae*

The symptoms appear in the form of brown discoloration of the flower buds during the elongation period of the inflorescence. The discoloration gradually spreads to the entire flower buds. The affected flower buds dry and are shed prematurely resulting in severe yield loss.

Control measures

Periodic collection and destruction of debris from the soil, proper maintenance of the vines by application of recommended doses of fertilizers & timely irrigation, cutting and removing the rotten shoots help to minimize the disease occurrence.

Spraying with Bordeaux mixture 1.0% or copper oxychloride 0.25% or mancozeb 0.2% or thiophanate methyl 0.1% or chlorothalonil 0.15% at the onset of the disease affords good control of the disease

General characters of genus *Colletotrichum* - Refer page - 136

v) Stem blight of vanilla - *Phytophthora meadii*

Stem blight of vanilla

'Stem blight' is a serious disease of vanilla and in severe cases may cause death of the vines. The symptoms appear on the stem as brown colored, blighted patches, sometimes extending several centimeters along the stem. The affected portions gradually shrink, the leaves turn yellow and subsequently the vines dry and die. The symptoms are more pronounced during the summer months.

vi) Bean rot of vanilla - *Phytophthora meadii*

'Bean rot' is another serious disease of vanilla that affects the beans of vanilla. The disease appears during the months of June - August when the Southwest monsoon rains are in full swing. Initially rotting symptoms develop at the tips of the beans, which slowly extend towards the pedicel resulting in the rotting of the whole bunch. Affected beans become soft, brown colored and show abundant mycelial growth. In later stages, the disease advances to the stem, leaves, aerial roots and extend towards the basal portion. The whole vine shows decaying symptoms and the entire vine perishes.

Bean rot of vanilla

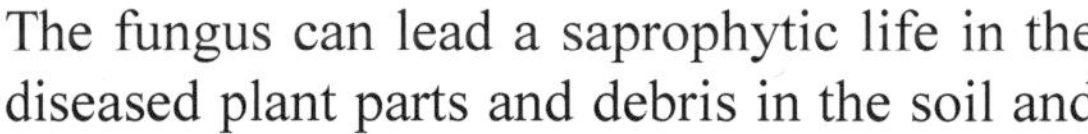

The fungus can lead a saprophytic life in the diseased plant parts and debris in the soil and continue to produce sporangia. The zoospores, which come out of the sporangia are disseminated by wind, rain splash and by some insects inadvertently and cause fresh infection. The oospores can remain in the soil for prolonged periods and produce zoospores when conditions become favorable and cause fresh infection.

Excess shade in the plantations, continuous rains, crowding of vines, poor aeration in the garden, water logged conditions are some of the factors that predispose the vines to infection, disease development and spread.

High relative humidity, rainfall and low temperatures between 20° - 25°C favor disease development.

Control measures

Periodical collection and destruction of disease affected plant debris from the plantations, removal and destruction of severely affected vines, provision of adequate drainage facilities to avoid water stagnation, removal of excessive shade so as to improve good aeration, proper maintenance of the vines by way of application of recommended doses of fertilizers, timely irrigation are the measures that can be implemented to prevent occurrence and spread of the disease.

Spraying with Bordeaux mixture 1.0% or copper oxychloride 0.25% or captafol 0.2% or mancozeb 0.2% or carbendazim 0.1% or a combination of mancozeb and carbendazim affords good control of the disease.

General characters of genus *Phytophthora* - Refer page - 142

vii) Vanilla necrosis - Poty Virus (VNPV)

'Vanilla necrosis' is a serious virus disease of vanilla and causes severe damage to the vines ultimately resulting in the death of the affected vines. The disease symptoms develop in the young leaves in the form of diffused chlorotic patches followed by leaf distortion. Necrotic lesions are produced on older leaves and stems subsequently. In advanced stages necrosis extends to large areas of the stem resulting in the death of the affected vines.

The virus is transmitted by the aphid vectors *Macrosiphum* spp. and *Myzus* spp. in a non-persistent manner. The virus can also spread by sap inoculation.

Control measures

Avoiding transportation of shoot cuttings or planting materials from diseased areas to disease-free areas, avoiding cuttings from affected vines for planting in new areas, uprooting and destroying virus affected vines, avoiding close planting and over crowding of vines in the gardens are some of the measures that can be followed to restrict the occurrence and spread of the disease.

The disease can be controlled to some extent by controlling the insect vectors by the application of suitable insecticides.

20. Basil *(Ocimum basilicum)*

Basil plants

'Basil' also known as 'Great basil' and 'Sweet basil' is an annual culinary herb of the family Lamiaceae. Basil is native to Tropical regions from Central Africa to Southeast Asia. It is a tender plant that grows up to a height of 150 cm. Its leaves are richly green and ovate. It has a thick, central tap root and the flowers, which grow from a central inflorescence are small and white-colored. The leaves of the plants, which are used in cousins world wide taste somewhat like that of anise with a strong pungent sweet smell. Sweet basil is used in many Mediterranean and Italian cousins. It adds a distinctive flavor to salads, pasta, pizza and many other dishes.

Sweet basil provides vitamins, minerals and a range of antioxidants. High concentration of Eugenol and Limonene present in basil have antioxidant properties. Basil provides health benefits in the diet, as herbal medicine and as an essential oil. It is traditionally used for the treatment of snake bites, colds and inflammation within the nasal passages. Basil reduces oxidative stress, supports liver health, fights cancer, protects skin aging, reduces high blood sugar, supports cardiovascular health, boosts mental health, reduces inflammation and swelling and combats infection.

Diseases of basil (*Ocimum basilicum*)

i) Gray mould of basil - *Botrytis cinerea*

Gray mould of basil

The most characteristic symptom of the disease is the appearance of a brown to gray, dense fungal growth on both leaves and stem. Diseased leaves die and eventually drop off from the plant. In case of severe infection of the stem region, the infected plant may die.

As a result of asexual reproduction the fungus produces abundant conidia, which are dispersed by wind and rain splash and continue to cause fresh infection. The mycelium can also survive in the diseased fallen debris in the soil and continue to produce conidia. The pathogen also produces highly resistant chlamydospores, which can remain in a viable state for long periods and perpetuate the disease.

High humidity, poor air circulation in the gardens, over crowding of plants and moderately cool temperatures are factors that favor disease occurrence and spread.

Control measures

Collection and destruction of plant debris from the soil, removal and destruction of leaves and severely affected plants, proper maintenance of the plants, avoiding over head irrigation, avoiding causing injuries to plants etc. are measures that can be adopted to ward off disease occurrence and spread.

Spray application of chlorothalonil 0.125% or mancozeb 0.2% or thiram 0.2% or wettable sulphur 0.4% has been found to be effective in controlling the disease.

General characters of genus *Botrytis* - Refer page - 176

ii) *Cercospora* leaf spot of basil - *Cercospora ocimicola*

Symptoms appear as small, scattered, circular to irregular, dark spots with light centers on the leaves.

Etiology, epidemiology, control measures and general characters of genus *Cercospora* - Refer leaf spot of fenugreek (Page 61)

Cercospora leaf spot of basil

Fusarium wilt of basil

iii) *Fusarium* wilt of basil - *Fusarium oxysporum* f.sp. *basilicum*

'***Fusarium* wilt'** is a very common disease of basil and has been reported from basil growing areas all over the world. The symptoms include general wilting of foliage, chlorosis and necrosis of leaves and apices, retardation of growth, asymmetric growth of the affected plants, dark streaks on the stems and petioles, stem necrosis, vascular discoloration, root rot and eventual death of the affected plants. After entry into the host, the fungus proliferates in the tissues and the mycelium and micro conidia produced in profusion invade the water conducting xylem vessels and blocks translocation of water and mineral nutrients to the foliage leading to wilting and subsequent death of the affected plants.

Etiology, epidemiology, control measures and general characters of genus *Fusarium* - Refer *Fusarium* wilt of coriander (Page 55)

iv) Downy mildew of basil - *Peronospora belbahrii*

Symptoms of the disease appear as yellow patches formed typically from around the middle vein and gradually spreading over the entire leaf surface. Simultaneously a fuzzy, greyish-purple, downy fungal growth develops on the under surface of the leaves. Soon the downy growth may cover large areas of the under surface of leaves. The downy growth comprises of large number of sporangiophores and sporangia, which are the asexual reproductive structures of the fungus. Eventually the leaves turn into brown to black, angular necrotic patches and the leaves die prematurely.

Downy mildew of basil

Etiology, epidemiology, control measures and general characters of genus *Peronospora* - Refer downy mildew of mustard (Page 82)

v) Damping off of basil - *Pythium ultimum*

Damping off of basil

The pathogen is particularly virulent on seedlings and causes root rot or damping off. It invades the host stem at or just below the soil surface and spreads rapidly producing a soft, dark brown rot. The affected seedlings collapse from the soil line or just above the soil line resulting in patches of collapsed and rotten seedlings.

Pythium ultimum is favored by cool temperatures below 20°C and high relative humidity and high levels of soil moisture.

Etiology, epidemiology, control measures and general characters of genus *Pythium* - Refer damping off of cardamom (Page 22)

vi) Bacterial leaf spot / Basil shoot blight - *Pseudomonas cichorii*

Basil shoot blight

Typical symptoms appear as water-soaked, brown & black spots on the leaves and streaking on the stem region. The leaf spots are angular or irregular and are limited by the small veins. Long distance distribution of the disease is through propagating materials carried from disease affected areas to areas where the disease is not prevalent. Local spread is mainly through rainwater splash and irrigation. Seeds, infected cuttings and infected seedlings are the main sources of infection.

High relative humidity, extended periods of leaf wetness, moderate temperatures ranging from 25° - 30°C, excessive shade and high levels of nitrogen are conducive for the occurrence and spread of the disease.

Control measures

There are no specific control measures for controlling the disease. Field sanitation measures, such as removal and destruction of diseased plant debris from the soil, removal and destruction of affected plants, avoiding overhead irrigation, judicious application of fertilizers, avoiding handling of plants when the leaves are wet, reducing humidity levels by improving air circulation, avoiding close planting etc. are measures that can be adopted to reduce the occurrence and severity of the disease.

21. Fennel - *Foeniculum vulgare*

'Fennel' is a flowering plant belonging to family - Apiaceae. It is indigenous to the shores of the Mediterranean, but now it is grown in many parts of the world. It is a hardy, erect, perennial herb with feathery leaves The plants grow up to a height of 2.5 meters. The flowers are produced in terminal compound umbels. The tiny yellow flowers are borne on short pedicels. The small dry fruits are brownish in color and the grooved seeds attached to the pericarp of the fruits. The volatile, essential oil present in the seeds, 'anethole' and 'fenchone' are responsible for imparting the sweet aroma.

Fennel plant

Fennel flowers

Fennel seeds

Leaf blight of fennel

Fennel as a spice is used as a delicious additive to fish & poultry dishes, cooked whole grains, pasta, sauces & many other vegetarian and non-vegetarian recipes.

Fennel is used for various digestive problems including heart burn, intestinal gas bloating, loss of appetite and colic problems.

Diseases of fennel (*Foeniculum vulgare*)

i) Leaf blight of fennel - *Ramularia foeniculi (Cercospora foeniculi)*

Initial symptoms appear as small, angular, brown, necrotic spots on the lower and older leaves. These spots enlarge and are covered with grayish-white, erumpent fungal growth. At later stages, linear, rectangular spots appear on the stem region, which gradually cover the entire stem, peduncles and fruits. Severely affected leaves shrivel and dry. In case of severe attack, the whole plant turns brown, wither and die.

Leaf blight of fennel

Etiology, epidemiology, control measures and general characters of genus *Cercospora* - Refer *Cercospora* leaf spot of fenugreek (Page 61)

ii) Leaf spot of fennel - *Cercosporidium punctum (Cercospora personata)*

The disease primarily attacks the lower, older leaves. Initial symptoms appear as small, circular or oval or irregular-shaped patches on the leaves and stems. Tips of affected leaves and stems turn brown to black in color and dry up. Affected leaves and stems show tiny, discrete, dark brown to black patches of fungal growth.

Etiology, epidemiology, control measures and general characters of genus *Cercospora* - Refer *Cercospora* leaf spot of fenugreek (Page 61)

iii) Damping off of fennel - *Pythium aphanidermatum*

'Damping off' occurs at the pre- and post-emergence stages. In the pre emergence phase, the seedlings are killed before they reach the soil surface. The young radical and plumule are killed and the seedlings rot completely. The post-emergence phase is characterized by the infection of the young, juvenile tissues of the collar region at the ground level and the infected tissues become soft, water-soaked and rot, as a result the seedlings topple over at the ground level and die in patches.

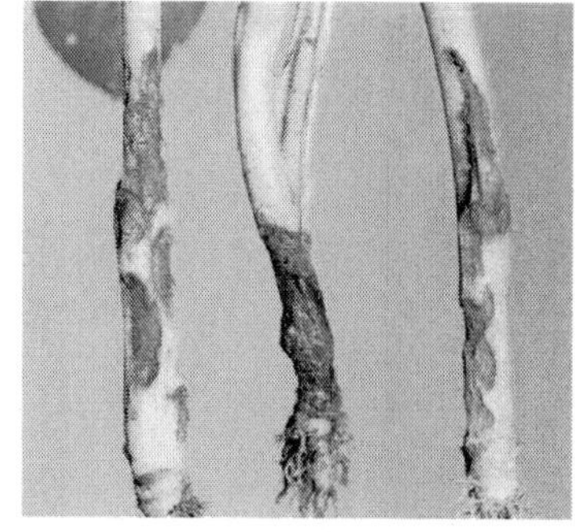

Damping off of fennel

Etiology, epidemiology, control measures and general characters of genus *Pythium* - Refer damping off of fenugreek (Page 67)

iv) Wilt of fennel - *Fusarium eqiseti*

Diseased plants show wilting symptoms at the seedling stage, as well as at later stages of plant growth. Infected plants turn yellow. Dark streaks are found in the vascular regions and fungal growth may be seen in the infected plant stem if split longitidinally.

Etiology, epidemiology, control measures and general characters of genus *Fusarium* - Refer *Fusarium* wilt of coriander (Page 55)

v) Powdery mildew of fennel - *Erysiphe polygoni*

Symptoms of the disease appear later in the growing season of the crop mostly at the flowering stage. The powdery fungal growth comprising of mycelium, conidiophores and conidia usually develop first on the leaves and later cover all succulent plant parts including leaves, petioles, stems, branches and even flowers. In severe cases of infection seed development may be completely arrested.

Powdery mildew of fennel

Etiology, epidemiology, control measures and general characters of genus *Erysiphe* - Refer Powdery mildew of coriander (Page 53)

vi) Collar rot of fennel - *Sclerotinia sclerotiorum*

The disease attacks the collar region of the plants and the affected plants decay at the collar region, turn yellow, topple over and eventually die.

Etiology, epidemiology, control measures and general characters of genus *Sclerotinia* - Refer stem rot of coriander (Page 57)

vii) Root rot of fennel - *Rhizoctonia solani*

Symptoms consist of seed decay and brown to reddish lesions on the stems and roots of emerged seedlings just below the soil line. On young plants reddish-brown, sunken lesions develop, which girdle the stems of affected plants resulting in their death. Affected plants appear stunted and unthrifty and die soon. The disease occurs often in patches in the field. On older plants the pathogen causes a reddish-brown, dry cortical root rot that may extend into the base of the stem region. Leaves of affected plants turn yellow and show wilting symptoms.

Etiology, epidemiology, control measures and general characters of genus *Rhizoctonia* - Refer blight of pepper (Page 18)

22. Saffron - *Crocus saffron*

'Saffron' commonly known as 'Saffron crocus' is a perennial, flowering plant species belonging to family - Iridaceae. The plant is indigenous to Greece to Southwest Asia. Presently it is cultivated on a large scale in Western Asia, Turkey, Iran, Greece, India and Spain. Iran accounts for about 90% of the world's production of saffron.

Saffron a delicate looking lavender plants grow up to a height of 10 - 30 cm. The plant has a corm, which holds the leaves and other aerial parts. The flowers are quite attractive and purple in color. Spice saffron is obtained from the filaments (crimson stigma and styles called threads) that grow inside the flowers. To obtain 450 g of saffron spice, 75,000 saffron blossoms are required. The volatile fraction 'Safranal' is responsible for saffron's distinctive aroma.

Saffron has a strong, exotic aroma and a bitter taste and is used to color and flavor many Mediterranean and Asian dishes particularly rice, fish and soups.

Saffron is widely used in Ayurvedic remedies for a wide range of ailments from Arthritis and Asthma to infertility and impotency. Saffron is known to strengthen the stomach and to promote its functioning.

Diseases of saffron (*Crocus sativus*)

i) Corm rot / Dry rot of saffron - *Fusarium moniliforme* var. *intermedium*, *Fusarium oxysporum* f.sp. *gladioli* and *Fusarium solani*

Symptoms of the disease appear as yellowing of leaves, which soon dry. The corms of such infected plants show brown, discolored patches that extend into the inner flesh of the corms. In severe cases discoloration extends to the top of the corm at the soil surface and into the nodal region of the stem. Mostly lesions develop near the base of the corm but may occur at any part of the corm and extend into the inner fleshy tissue. In later stages the lesions become deeply sunken forming dark patches that are hard and rough. Roots from the affected corms also become brown and decayed. Severely affected corms rot completely and eventually the plants die.

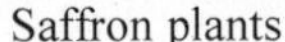

Saffron plants

Saffron flowers

Corms of saffron

Saffron

Etiology, epidemiology, control measures and general characters of genus *Fusarium* - Refer *Fusarium* yellows of ginger (Page 7)

ii) Bulb/corm rot of saffron - *Sclerotium rolfsi*

Symptoms of the disease appear as brown to dark brown, sunken, irregular patches below the corm scales. The lesions are sunken with raised margins. Infected corms show drying of leaves from tips downward. Whitish fungal growth appears on the infected corms and the corms rot leading to death of the infected plants ultimately. In the late stage of disease development sclerotia are formed in large numbers on the fungal growth

Etiology, epidemiology, control measures and general characters of genus *Sclerotium* - Refer Basal wilt of pepper (Page 18)

Corm rot of saffron

23. Chicory - *Cichorium intybus*

'Chicory' also known as **'Coffee weed'**, is a somewhat woody perennial herbaceous plant belonging to the family - Asteraceae. It is a native of the Mediterranean region of Europe. Now it is being cultivated in many countries of the world particularly in North America, China and Australia.

Chicory plants

Chicory roots

Dried chicory roots

Chicory plants grow up to a height of 100 cm. It has lanceolate and irregularly serrated leaves. It produces bright, sky blue, sometimes pink or white-colored flowers. Flowers are produced in clusters at the tips of branches. The fruits or seeds are oblong, with angled edges and blunt ends with a fringe of short bristles at the wider end. The plants have relatively large, brown, fleshy, branched tap root. The roots are boiled, roasted, ground and used as an additive with coffee in various proportions.

Chicory aids in bowel movements, improves blood sugar control and supports weight loss besides many other ailments.

Diseases of chicory (*Cichorium intybus*)

i) Downy mildew of chicory - *Bremia lactucae*

Downy mildew of chicory

The obligate fungus can infect chicory plants at any stage of growth. The first symptoms of downy mildew appear as chlorotic, yellow spots or patches on the upper leaf surface. Within one or two days, a white, fluffy fungal growth appears on the under surface of the leaf. This growth indicates the formation of sporangiophores and sporangia, which are the asexual reproductive structures. The lesions enlarge and become more chlorotic and eventually cover the entire leaf surface. Severely affected leaves turn completely brown and die.

The disease is spread by sporangia, which are carried over long distances by wind. The oospores present in the diseased crop debris in the soil cause primary infection on young plants. The sporangia produced in the alternate host lettuce may also cause infection of chicory plants growing nearby.

Moist and cool weather conditions are favorable for occurrence and spread of the disease. Relatively low night temperatures and dew fall are conducive for the germination of sporangia and infection.

Control measures

Removal and destruction of diseased crop debris from the soil, removal of severely affected plants, avoiding cultivation of lettuce nearby chicory fields, using disease-free seeds for sowing and other field sanitation measures may help to prevent occurrence and spread of the disease.

Foliar spraying with mancozeb 0.2% or metalaxyl 0.1% as soon as initial symptoms of the disease is noticed is effective in controlling the disease Spraying a mixture of mancozeb 0.2% and metalaxyl 0.1% has been found to be very effective in controlling the disease.

General characters of genus *Bremia* - Refer page - 178

ii) *Fusarium* wilt of chicory - *Fusarium oxysporum* f.sp. *cichorii*

Symptoms of the disease appear as small, water-soaked lesions on the foliage followed by drooping and yellowing of all the leaves. The affected plants die soon after. Whitish mycelial growth and pinkish, slimy masses of conidia can be seen on the stem region near the soil level.

Etiology, epidemiology, control measures and general characters of genus *Fusarium* - Refer *Fusarium* wilt of coriander (Page 55)

iii) *Septoria* blight of chicory - *Septoria lactucae*

Initial symptoms appear as small, irregularly shaped, yellowish, scattered spots on older leaves. The spots gradually enlarge, turn pale brown and finally dark brown in color. The leaf spots may be surrounded by a yellow halo. As the disease advances, the centre of the spots turn grey with a distinct dark margin and the central areas of old lesions may fall off leaving shot holes. In some cases the lesions may coalesce and form large necrotic patches. Eventually the affected leaves wilt, dry and die. Usually the older leaves are affected first and then the disease gradually spreads to the upper younger leaves. The disease attacks the stem regions also and develops linear, dark brown lesions. Minute, black, pin point-like dots, which are the pycnidia appear on the lesions.

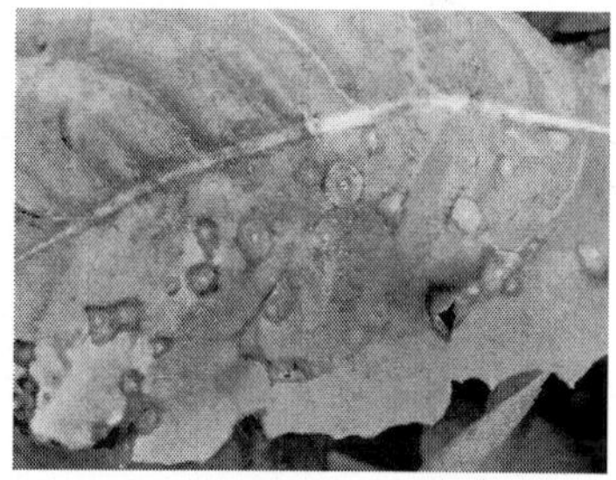

Septoria blight of chicory

The pycnidiospores produced from the pycnidia are spread by strong winds, rain splash, irrigation water and agricultural implements and cattle along with soil. The fungus over seasons as mycelium and as pycnidiospores within the pycnidia on the diseased plant debris in the soil. The fungus may survive in the plant debris for up to 2 years. The pathogen may also be carried through infected seeds. Conidia produced from alternate hosts may also initiate fresh infection on chicory.

The fungus requires high moisture for infection and disease development, but can cause disease at a wide range of temperatures from 10° - 27°C. The disease incidence is severe during the rainy seasons. Long periods of rainfall, persistent dew and high atmospheric humidity favor disease occurrence and spread.

Control measure

Selection of disease-free seeds, removal and destruction of diseased crop debris from the soil, removal of severely affected plants, avoiding cultivation of alternate hosts nearby chicory fields, crop rotation and other field sanitation measures may help to prevent occurrence and spread of the disease.

Spraying the crop with Bordeaux mixture 1.0% or copper oxychloride 0.25% or mancozeb 0.2% or captan 0.15% from the time initial symptoms of the disease are noticed affords good control of the disease.

iv) White mould of chicory - *Sclerotinia sclerotiorum* f.sp. *carotovora*

Symptoms of the disease appear as abundant whitish mycelial growth (thread-like fungal structures) on any part of the plant and wilting of the outer, older leaves. The wilting spreads inwards until all the leaves wilt and the affected plant turns yellow. Soft, watery lesions may develop on the leaves. Eventually all the leaves collapse and drop off. Black fungal structures, which are the sclerotia are formed on the infected leaf tissues and on the soil surface.

Etiology, epidemiology, control measures and general characters of genus Sclerotium - Refer stem rot of coriander (Page 57)

v) Bottom rot of chicory - *Rhizoctonia solani*

Small, red to brown spots appear on the lower, older leaves usually on the under surface of mid ribs. The spots enlarge rapidly causing the leaves to rot. Amber-colored liquid may ooze out from the leaf lesions. Fungal growth may develop on the root lesions and black sclerotia may be formed in the mycelial mat.

Etiology, epidemiology, control measures and general characters of genus *Rhizoctonia* - Refer blight of pepper (Page 17)

vi) Bacterial soft rot of roots of chicory - *Erwinia carotovora*

Yellow discolored plants, wilted foliage and black necrosis of petioles near the crown are the characteristic symptoms of the disease. Wilted leaves subsequently collapse and die forming a dry, brown to black rosette. The root and crown rot partially or wholly and the rotten tissues become soft and pulpy. Affected areas of the root appear as slimy masses, which turn brown or black eventually.

The bacteria causing soft rot can survive in the diseased plant debris and continue to multiply. The disease spreads through bacteria carried along with soil by tools used for cultural operations and by irrigation water. The bacteria enter the plants through wounds. They are anaerobic, rod-shaped, occurring singly or in pairs, non-spore forming, encapsulated and motile by means of 2 - 8 peritrichous flagella.

Warm, humid conditions favor infection and development of the disease.

Control measures

Field sanitation measures, such as removal and destruction of diseased plant debris from the soil, removal and destruction of severely affected plants, provision of suitable drainage facilities, proper maintenance of the crop, judicious application of recommended doses of fertilizers, taking care not to cause injuries to the plants, application of suitable insecticides to eliminate insect pests that may cause injuries to the plants are some of the measures that can be followed to ward off disease occurrence and spread

Drenching the soil around the plants with Bordeaux mixture 1.0% or copper oxychloride 0.25% affords fairly good control of the disease.

24. Curry leaf - *Murraya koenigii*

'Curry leaf tree' otherwise known as **'Sweet neem'** is a tropical to subtropical tree in the family Rutaceae and is native to Asia. It is a small tree growing up to a height up to about 6.0 metres. The aromatic leaves are pinnate with 11 - 21 leaflets. The plants produce small, white flowers and small, shiny-black drupes containing a single large seed. The berry pulp is edible with a sweet flavor.

Curry leaf tree

Curry leaves are commonly added to dishes to bring about a robust, rich flavor and popularly used in meat dishes, curries and other traditional Indian recipes. Curry leaves are often garnished in oil or butter before the fat and the cooked leaves are added to dishes.

Curry leaves offer antibacterial, antidiabetic, pain relieving and anti-inflammatory effects. Curry leaves are rich in protective plant substances, such as alkaloids, glycosides and phenolic compounds, that give the fragrant herb potent health benefits.

Diseases of curry leaf (*Murraya koenigii)*

i) *Phyllosticta* leaf spot of curry leaf - *Phyllosticta murrayae*

The pathogen produces small, scattered, oval to elongated, water-soaked spots in large numbers on the leaves. The spots are distributed irregularly over the entire surface of leaves. The brown spots have a dark brown margin and are surrounded by a yellow halo. The disease leads to large scale defoliation when the disease incidence is severe.

Etiology, epidemiology, control measures and general characters of genus *Phyllosticta* - Refer *Phyllosticta* leaf spot of ginger (Page 6)

Phyllosticta leaf spot of curry leaf

ii) *Colletotrichum* leaf spot of curry leaf - *Colletotrichum gloeosporioides*

Symptoms of the disease appear as small, black, circular to oval, necrotic spots on the leaves. The spots may enlarge to form bigger spots. Severely affected leaves drop off prematurely.

Etiology, epidemiology, control measures and general characters of genus *Colletotrichum*- Refer *Colletotrichum* leaf spot of ginger (Page 7)

25. Lemon grass - *Cymbopogon citritus*

Lemon grass

'Lemon grass' also called as **'Citronella grass'** is a perennial grass in the family Poaceae, grown for its fragrant leaves and stalks, which are used as flavoring agents. Lemon grass is native to Sri Lanka or South India or Malaysia. Now it is grown in many countries around the world. The grass grows in dense clumps and has several stiff stems and slender, blade-like leaves. The plants grow up to a height of 1.8 metres and their life span is about 4.0 years. The leaves are blue-green in color, which turn red at later stages and emit a strong lemon fragrance. Lemon grass produces large compound flowers on spikes. The stalks of the plants are commonly used to flavor many dishes in Southeast Asian cooking. The heart of young shoots are cooked and consumed as a vegetable. The leaves are used to flavor various vegetarian and non-vegetarian dishes. Leaves are also widely used to make lemon grass tea. The essential oil extracted from lemon grass contains citral, borneol, estragole, methyl eugenol and geranyl acetate. The oil is used commonly in insect repellants, perfumes and soaps.

Lemon grass tea helps in relieving anxiety, lowering cholesterol, preventing infection, boosting oral health, relieving pain, boosting red blood cell levels and relieving bloating. Lemon grass tea is also a delicious drink.

Disease of lemon grass (*Cymbopogon citratus*)

i) Rust disease of lemon grass - *Puccinia nakarishikii*

Rust disease of lemon grass

The disease appears first on the lower, older leaves and then spreads to the other leaves. Symptoms of the disease appear as tiny, yellow spots that develop into brown spots and elongated stripe-like, brown lesions that run parallel to the leaf veins. Lesions develop on both the upper and lower surfaces of leaves. Only the uredinial and telial stages are formed. Pustules of uredinia are lighter in color than the telial pustules. Lesions may enlarge and coalesce forming large leaf spots or blights. Heavily infected leaves become discolored and necrotic in streaked patterns and die prematurely. Only the uredinial pustules on the lower surface of leaves erupt and release urediniospores in large numbers, which are dispersed by wind.

Etiology, epidemiology, control measures and general characters of genus *Puccinia* - Refer rust of allspice (Page 96)

ii) Red leaf spot of lemon grass - *Colletotrichum graminicola*

Symptoms appear in the form of small, brown spots on the under surface of leaves. The spots enlarge and become bigger patches with concentric rings in the central portion. Spots may be formed on the leaf sheaths and mid ribs. Affected leaves turn completely brown, dry and die.

Etiology, epidemiology, control measures and general characters of genus *Colletotrichum* - Refer leaf spot of ginger (Page 7)

iii) Leaf blight of lemon grass - *Curvularia andropogonis*

Older leaves are more prone to attack by the disease. Symptoms appear in the form of minute, circular, reddish-brown spots mostly on the margins and tips of leaves. These spots coalesce to form elongated, reddish-brown, necrotic lesions. Severely affected leaves turn completely brown, dry and die prematurely.

The fungus can lead a saprophytic life in the infected plant debris lying in the soil as mycelium and continue to produce conidia and cause primary infection. Secondary infection is by wind-borne conidia and by other agents like irrigation water and soil.

Warm and humid conditions favor disease occurrence and spread.

Control measures

Field sanitation measures, such as removal and destruction of diseased plant debris from the soil, removal and destruction of severely affected plants, provision of suitable drainage facilities, proper maintenance of the crop, judicious application of recommended doses of fertilizers and other such agronomical practices help to reduce the occurrence and spread of the disease.

Spraying the crop with Bordeaux mixture 1.0% or copper oxychloride 0.25% or mancozeb 0.2% affords good control of the disease.

General characters of genus *Curvularia*

26. Tamarind - *Tamarindus indica*

'Tamarind' a leguminous tree belonging to family - Fabaceae is indigenous to Tropical Africa. Tamarinds are long-lived, ever green, massive trees that can grow up to a height of 24 - 30 meters. The leaves are bright green, pinnate and feathery in appearance. The flowers are borne in small racemes and are yellow with orange or red streaks. The pod-like fruits produced are long, brown and are irregularly curved. When fully ripe, the shells become brittle and are easily broken. The pulp dehydrates to a sticky paste enclosed by a few coarse strands of fiber. The pods may contain 1 - 12 large, flattish, glossy, brown, hard seeds embedded in the brown edible pulp. The pulp has a pleasing, tangy, sweet and sour flavor and is high in both acid and sugar. It is also rich in vitamin B and calcium.

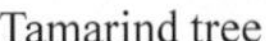

Tamarind tree

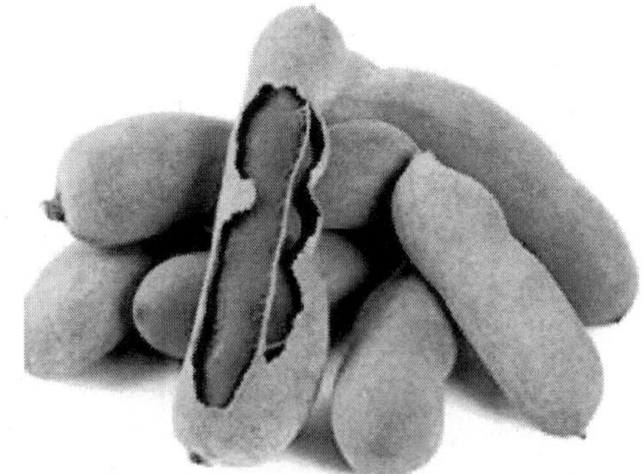

Tamarind fruit

Tamarind pulp is used in several cuisines around the world. Tamarind paste is used in chutnies and curries, especially in fish curries. The pulp is also used in many traditional medicines and as a metal polish. The wood of the tree can be used for wood working. Tamarind oil exacted from the seeds is used in many industries. Tender, young leaves of tamarind are used in Indian cuisines. Because tamarind has multiple uses, it is cultivated around the world in tropical and sub tropical zones.

Diseases of tamarind *(Tamarindus indica)*

i) Powdery mildew of tamarind - *Erysiphe polygoni*

Powdery mildew of tamarind

Symptoms of the disease appear as small, whitish, circular or irregular, powdery patches on both the surfaces of leaflets. The powdery patches enlarge rapidly and often cover the entire area of the leaflets. The leaves appear as if covered by a grayish-white powdery coating from a distance. The powdery growth consists of the mycelium, conidiophores and conidia, which are produced in profusion. Eventually the leaflets turn yellow completely, dry and drop off. Minute, black, pin point-like dots representing the sexual fruiting bodies, cleistothecia are found embedded in the mycelial mat.

Etiology, epidemiology, control measures and general characters of genus *Erysiphe* - Refer powdery mildew of coriander (Page 53)

ii) Leaf spot of tamarind - *Phyllosticta tamarindicola*

The pathogen produces small, brown, scattered, oval to elongated, water-soaked spots in large numbers on the leaflets. The spots are distributed irregularly over the entire surface of the leaflets. The brown spots have a dark brown margin and are surrounded by a yellow halo. Minute, dark brown to black, pin point-like dots representing the pycnidia, which are the asexual fruiting bodies of the fungus are seen on the central area of the spots. Severely affected leaflets turn yellow completely and drop off prematurely.

Etiology, epidemiology, control measures and general characters of genus *Phyllosticta* - Refer *Phyllosticta* leaf spot of ginger (Page 6)

27. Red chilli / Cayenne pepper - *Capsicum annuum*

'Chilli pepper' *(Capsicum annuum)* plants, which belongs to family Solanaceae are perennial in nature with varying structures. They can be herbaceous or shrub-like, but are generally branching with greenish-brown stems and simple oval, bright green leaves. The plants produce flowers with five petals, which are usually white in color. Chilli pepper plants can grow to a height of 0.5 - 1.5

meters. They are commonly grown as annuals and the fruits are harvested in one growing season. *Capsicum annuum* species are believed to have originated from Mexico, while the others are all likely to have originated from South America. Now chilli is grown all over the world and India is one of the major chilli producing countries of the world. The substances that give chilli pepper their pungency (spicy heat) are 'Capsaicin' and several other related chemicals called 'Capsaicinoids'

Red chilli plants

Red chilli fruits

Chilli peppers are widely used as a spice in many cuisines all over the world to add heat to dishes. Chilli pods are dried and then ground into chilli powder that is used as a spice. Chilli powder plays a very important role in Indian cooking. Besides culinary uses, chilli has many medicinal uses.

Capsicum is helpful in preventing coronary heart disease. *Capsicum* is both warming and vasodilative. It is also an antioxidant agent that allows the phytochemicals to circulate throughout the body and protects and repairs tissues against DNA damage. As an immunity booster, it helps in repairing damaged brain tissues and lowering the risk of oxidative stress, paediatric asthma and cancer. It also improves bone health. Vitamins present in *Capsicum* have an antioxidant effect on cell tissues, which also improves skin health and delay ageing. For women it works best in relieving menopausal symptoms.

Diseases of red chilli (*Capsicum annuum*)

i) Damping off of seedlings of red chilli - *Pythium aphanidermatun* and *P. de baryanum*

'Damping off of seedlings' is very common all over the world. The disease occurs in all kinds of soils both under tropical and temperate climates all through the year. Pre-emergence and post-emergence damping off may occur. In pre-emergence damping off, the seed and the radical rot before the seedling emerges from the soil, while in post-emergence damping off, which is more conspicuous than pre-emergence damping off, the newly emerged seedling is killed after its emergence from the soil. Stems of affected seedlings rot at the ground level and they collapse, topple over and die. Infection usually occurs at or just below ground level before the stem has hardened sufficiently to resist attack by the pathogen. The tissue in the infected area becomes soft and water-soaked. As the disease advances, the stem becomes constricted at the base, turns dark and the seedling topples over, rots and dies. In the nursery beds the symptoms develop suddenly, killing large number of seedlings in patches around the foci of infection.

Damping off of seedlings of red chilli

The damping off fungi are natural inhabitants of soil. They survive on dead plant and animal wastes and live as saprophytes or as parasites on fibrous roots of plants and cause primary infection. The oospores produced by these fungi are resistant to adverse soil temperature and moisture conditions, remain dormant in the soil for prolonged periods and are the most important source of primary infection. Secondary infection is by sporangia produced by the fungi.

Temperatures between 10° - 18°C induce germination of sporangia and oospores by means of zoospores, while temperatures above 18°C favor germination by germ tubes. In soils with high soil moisture content and in ill-drained, ill-aerated and poor, hard clay soils the incidence of the disease is more severe.

Control measures

Selection of nursery areas in elevated places with good drainage facilities, establishing raised nursery beds, avoiding thick sowing, light and frequent irrigation and other field sanitation measures help to ward off occurrence of the disease.

Seed treatment with captan or thiram at 4.0 g or metalaxyl at 2.0 g per kg of seeds affords protection from pre-emergence damping off. Drenching the soil with captan 0.2% or thiram 0.2% or copper oxychloride 0.25% after sowing followed by another drenching after emergence of the seedlings protects the seedlings from damping off disease.

Seed treatment with spores of *Trichoderma harzianum* or *Penicillium oxalicum* or cells of *Pseudomonas fluorescens* may serve as effective biocontrol agents

General characters of genus *Pythium* - Refer page - 133

ii) Die-back and Fruit rot of red chilli - *Colletotrichum capsici*

It is a serious disease of chilli crop and occurs in all chilli growing countries of the world. The disease is characterized by necrosis of tender twigs from the tip downwards. The symptoms may appear in a single branch or in a few branches or the entire plant may be affected and start drying from the tips. The necrotic twigs are water-soaked and turn brown in color. Within a few days the necrotic areas turn grayish-white or straw-colored. On these areas numerous, black, scattered, pin point-like dots, which represent acervuli, the asexual fruiting bodies of the fungus are found. Sometimes the necrotic areas are demarcated from the healthy areas by a dark brown to black, irregular band. The fruit production in the infected plants are badly affected.

Die back of chilli

Fruit rot of chilli

The pathogen also attacks the ripening fruits, which start turning red. Initially one or a few small, black, circular spots appear on the skin of the fruits. The spots enlarge gradually in the direction of the long axis of the fruits and become somewhat elliptical. The spots, which are slightly sunken get diffused and become dirty gray or they may be surrounded by a thick and sharp, black margin, while the central area becomes grayish black or straw-colored. Sometimes the entire surface of the fruit turns straw-colored or dirty white from its normal red color. On such discolored areas, numerous, black, scattered dots, representing the acervuli are formed, often in concentric zones. The inner surface of the skin is covered with whitish-gray mycelial growth. The seeds are also covered by a mat of mycelium and are rust-colored. Affected fruits drop off easily. The disease may affect the ripe fruits in storage also.

The pathogen survives in the diseased plant debris in the soil for prolonged periods, sometimes even up to 12 months and continues to produce conidia. Seeds from infected fruits also carry the primary inoculum. Secondary spread of the disease is mainly through wind-borne conidia.

The disease incidence is more after the rainy season, followed by prolonged periods of dew fall. Temperatures ranging from 28° - 30°C and relative humidity above 95% and dew fall are favorable for spore germination and disease development. The pathogen has many alternate hosts, such as coriander, turmeric, marigold, pothos etc. and the spores produced from such hosts may also cause fresh infection.

Control measures

Selection of disease-free seeds, removal and destruction of diseased plant debris from the soil, removal and destruction of affected branches, prevention of alternate hosts from the vicinity of chilli fields, sorting out diseased fruits from healthy fruits in storage and other such field sanitation methods help to ward off the disease occurrence.

Spraying the crop with mancozeb 0.2% or copper oxychloride 0.25% or wettable sulphur 0.4% affords control of the disease

Seed treatment with captan or thiram or zineb at 4.0 g per kg of seeds is effective in eliminating the externally seed-borne inoculum.

General characters of genus *Colletotrichum* - Refer page - 136

iii) *Alternaria* leaf spot of red chilli - *Alternaria solani*

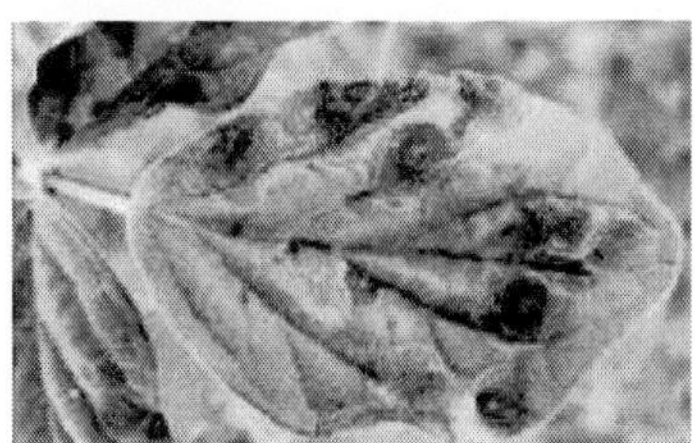

Alternaria leaf spot of red chilli

Initial symptoms of the disease appear as small, isolated, brown spots, scattered on the leaves. The older leaves are affected first and then the disease spreads to the upper, younger leaves. As the spots enlarge, the central tissues become necrotic and concentric rings are formed as in a target board, which is characteristic of the disease. A narrow chlorotic zone surrounds the necrotic area. The chlorotic zone is formed due to the production of a toxin, 'alternaric acid', which moves ahead of the necrotic area. As the spots increase in size, the chlorotic zone also expands. A few to many spots may develop in a leaf and occupy major portion of the lamina. In dry weather, the spots become hard and shrink and the leaves curl. When the weather is moist, the spots are covered with a dense, greenish-blue growth of the fungus. The spots coalesce and become rotten patches. In case of severe attack, the leaves shrivel, dry and fall down prematurely.

The mycelium of the fungus remains viable in dry, infected leaves for a year or more. The conidia have also been found to be in a viable state in the infected plant debris in the soil for about 17 months and cause fresh infection. The lower leaves near the ground level are infected first through conidia found in the soil. Secondary spread occurs through conidia produced on the primary spots. Wind, water and some insects disperse the conidia. The conidia produced from alternate hosts also serve as primary inoculum.

High atmospheric humidity, frequent rains followed by warm and dry weather conditions are conducive for the rapid development and spread of the disease.

The optimum temperature range for germination of conidia is 28° - 30°C and at these temperatures the conidia germinate in 35 - 45 minutes.

General characters of genus *Alternaria* - Refer page - 153

iv) Powdery mildew of red chilli - *Leveillula taurica*

Initial symptoms of the disease appear as a white, powdery fungal growth on the under surface of leaves with light green to yellow blotches on the upper leaf surface. These areas turn brown with time and the affected areas coalesce causing a general yellowing of the entire leaf. The outer edges of the leaves may curl upward. The older leaves near the bottom of the canopy are usually infected first and exhibit symptoms before the younger leaves.

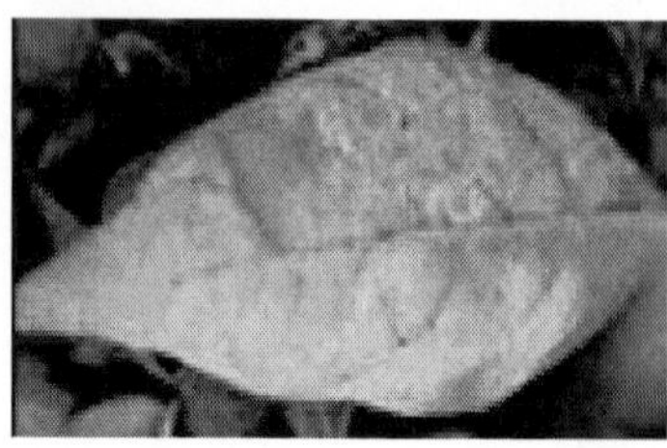

Powdery mildew of red chilli

Etiology, epidemiology, control measures and genera characters of genus *Leveillula* - Refer powdery mildew of garlic (Page 51)

v) *Phytophthora* blight of red chilli - *Phytophthora capsici*

The disease is more common in wet, water-logged areas. The fungus can invade all plant parts causing different disease syndromes viz., leaf blight, fruit rot and root rot. Affected plants wilt and die leaving brown stalks and leaves and small, poor quality fruits. When roots are affected, the plants cannot obtain enough water and nutrients from the soil due to rotting of the roots as a result the plants suddenly wilt and eventually die. Leaf symptoms appear as brown or black spots that may kill a localized portion of the plant. Affected areas are often bordered with a white moldy fungal growth.

Phytophthora blight

Etiology, epidemiology, control measures & genera characters of genus *Phytophthora* - Refer foot rot of pepper (Page 13)

vi) *Cercospora* leaf spot / Frog-eye leaf spot of red chilli - *Cercospora capsici*

The disease is characterized by the formation of small, reddish-brown, circular leaf lesions that have a watery appearance. The spots gradually enlarge and the center of the spots turns ashy-white, surrounded by a brown border. Later the center becomes white and necrotic. A typical spot appears white in the center, surrounded by a gray or brown ring, which is surrounded by a dark-brown or black ring and appears as a frog's eye. Several such spots may develop on a single leaf. The spots at the tip and margins of the leaf coalesce, forming big patches and the leaf starts drying inwards and the affected leaf dries completely.

Frog-eye leaf spot of red chilli

Etiology, epidemiology, control measures and general characters of genus *Cercospora* - Refer *Cercospora* leaf spot of fenugreek (Page 61)

vii) *Fusarium* wilt of red chilli - *Fusarium oxysporum* f.sp. *capsici*

The disease is characterized by sudden wilting of the affected plants and upward and inward rolling of the leaves. The leaves turn completely yellow and die prematurely.

Fusarium wilt of red chilli

By the time above-ground symptoms become evident, the vascular system of the plant is discolored particularly in the lower part of the stem and roots. The invading pathogen grows profusely inside the tissues and produces conidia in large numbers. The mycelium and the spores block the xylem vessels as a result the tissues rot and the tissues are discolored. The mechanical blocking of the conducting vessels obstructs the uptake of water and minerals from the soil to the above ground parts of the plants resulting in the sudden wilting of the affected plants.

Etiology, epidemiology, control measures and general characters of genus *Fusarium* – Refer *Fusarium* wilt of coriander (Page 55)

viii) Bacterial leaf spot of red chilli - *Xanthomonas campestris* p.v. *vesicatoria*

Initial symptoms appear as small, dark brown or black, circular or irregularly circular, water-soaked spots on the leaves. As the spots enlarge in size, the center becomes lighter and the spots are surrounded by a black margin. The spots may coalesce to form larger, irregular spots. Severely affected leaves turn yellow, wilt and drop off prematurely. Linear, dark brown to black lesions may develop on the petioles and stem region. Stem infection may often lead to cankerous growth and the branches wilt and die. On the green fruits small, scattered, circular, black, water-soaked spots with a light yellow border are formed. Slimy droplets of bacterial ooze come out from the spots. The affected fruits may drop off prematurely.

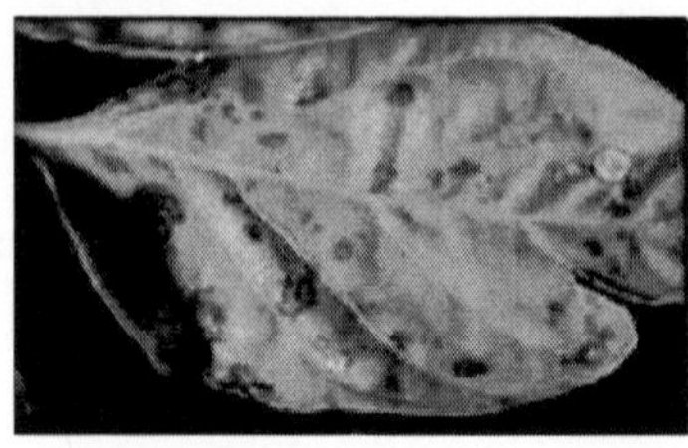

Bacterial leaf spot of red chilli

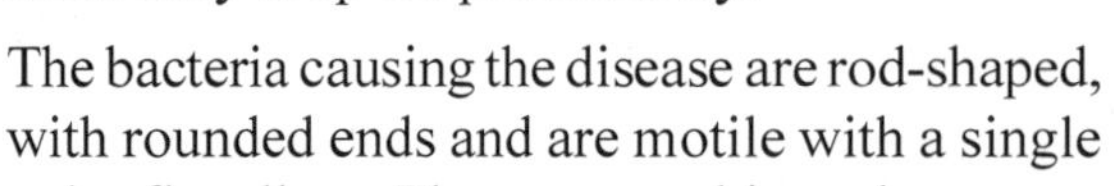

The bacteria causing the disease are rod-shaped, with rounded ends and are motile with a single polar flagellum. They are aerobic and are non-spore forming.

The bacteria can survive in the diseased crop debris and fallen diseased fruits in the soil for a very long time. They are also carried through seeds. The bacteria infect the germinating seedlings through the stomata on the cotyledons and cause primary infection. The bacteria may also gain entry into the leaves through the stomata, hydathodes, injuries caused by insects and mechanical means. Infected seedlings also carry the disease to the main field. Secondary spread is caused by rain splash, strong winds and by some insects accidentally.

Plants at the age of 40 - 50 days are more prone to attack by the bacteria in the main field. Moderate temperatures ranging from 24° - 28°C, high atmospheric humidity and intermittent rainfall favor infection and development of the disease. The bacteria also attack tomato and a few other species of *Capsicum.*

Control measures

Selection of seeds from disease-free, good quality fruits, selection of disease-free, healthy seedlings for transplanting, setting up of nurseries in disease-free areas, removal of alternate hosts from the vicinity of chilli fields, proper maintenance of the crop by supplying recommended doses of fertilizers, providing adequate supply of irrigation water and adopting other field sanitation measures help to prevent occurrence and spread of the disease.

Seed treatment with captan or thiram at 4.0 g / kg of seed or hot water treatment of seeds at 50°C for a continuous period of 25 minutes eliminates seed-borne inoculum.

Spraying with Agrimycin 100 at the rate of 800 - 1000 ppm (0.8 - 1.0 g / liter of water) or Streptocycline at 20 ppm (0.2 g / liter of water) at fortnightly intervals during the monsoon periods is found to be effective in controlling the disease.

ix) Mosaic disease of red chilli - *Tobacco mosaic virus / Tobacco virus* - 1

The disease occurs at all growth stages of the crop. In the early stages of infection, the interveinal regions become pale green in color. This is followed by formation of light- and dark-green, irregular, indistinct patches. Usually the areas adjacent to the veins remain green. Dark green blisters and enations may appear

Mosaic disease of red chilli

on the lower surface of leaves due to hyperplasia and hypertrophy of cells resulting in crinkling of the leaves. Necrotic, dark-brown spots or scorched patches known as 'mosaic scorch' or 'mosaic burn' may appear during hot spells. Affected plants are very much stunted. When disease intensity is high the leaves become thin, puckered, distorted, malformed and narrow with elongated central veins, giving the appearance of 'rat tail' to the leaves. The disease also causes partial sterility.

The virus causing the disease is single, rigid, rod-shaped, helical particle and contains a single molecule of ssRNA with a distinct central hole. The protein capsid functions as a protective covering around the RNA.

The virus is sap transmissible and highly contagious. It enters the host through any kind of injuries caused by mechanical means or by insects. It is not transmitted by insect vectors or through seeds. The virus is capable of withstanding highly adverse environmental conditions and remains virulent for a long time and can cause infection. Natural forces, such as strong wind, lashing rain etc., which make the leaves to rub against one another cause invisible injuries through which the virus gains entry into the host. Touching a diseased leaf and then touching a healthy leaf can transmit the virus. Cultural operations, such as weeding, intercultural operations, which are usually followed result in wounds that serve as portholes of entry for the virus into the host. Over 116 plant species belonging to 29 families, such as tobacco, tomato, beans, eggplant, *Datura* etc. serve as alternate host for the virus.

Control measures

Removing and destroying diseased plant debris in the soil, avoiding planting of disease affected seedlings, washing hands frequently with soap and running water during field operations, such as planting, weeding etc., uprooting and destroying disease affected plants in the nursery as well as in the main field, avoiding handling of healthy plants after handling diseased plants, eliminating solanaceous weed hosts in and around chilli fields, avoiding cultivation of alternate host crops in the vicinity of chilli fields, adopting wider spacing between plants, following crop rotation practices, growing resistant varieties and other field sanitation methods help to ward off disease occurrence and spread

Leaf extracts of some plant species, such as *Agave americana, Bougainvillea spectabilis, Clerodendron fragrans, Azadirachta indica, Thevetia neriifolia, Carica papaya* etc., inhibit the virus. Prophylactic spraying with 1.0% leaf extract of any one of the above plants reduces the incidence of the disease.

28. Green chilli / Chilli pepper - *Capsicum* spp.

All the diseases that attack red chilli *(Capsicum annuum)* attack all other species of *Capsicum* including green chilli.

Green chilli plant

Green chilli fruits

29. Asafoetida (*Ferula asafoetida*)

'Asaphoetida', 'Perumkayam' is an important spice used in various Indian recipes. It adds flavor to the food and is beneficial for health too. Asafoetida is the dried aromatic gum resin (Oleoresin) exuded from the living rhizome, root stock or tap root of several plant species of the Genus - *Ferula* (*Ferula foetida, F. narthex and F. assa-foetida*). The plants are monoecious, perennial herbs belonging to the Family - Umbelliferae. The plants are native to the deserts of Iran and mountains of Afghanistan. They are mainly cultivated in Pakistan, India, Kashmir, Tibet and Afghanistan

Asafoetida plants

Asafoetida resin

Asafoetida has a pungent smell and is generally called 'Stinking gum'. It is also known as 'Food of Gods' and 'Devils dung'. In cooked dishes it gives a smooth and pleasant flavor. It is mainly used as a flavoring agent and forms a constituent of many spice mixtures. Asafoetida has several medicinal properties. It is antiviral, antibacterial, antifungal and antioxidant in nature. The gum resin has antispasmodic, carminative, expectorant, laxative and sedative properties. Asafoetida is useful in the treatment of respiratory disorders, whooping cough, asthma, bronchitis, ,blood pressure and toothache. It is very effective as a drug that expels accumulated gases from the stomach and counteracts any spasmodic disorders. It is also a nerve stimulant and digestive agent. Antioxidant compounds, such as 'Ferulic acid' and 'Umbelliprenin' present in asaphoetida help to fight and reduce cancerous cells

Asafoetida is used in products meant to repel dogs, cats and wild life. It is also used as an insect attractant in light traps.

Diseases of asafoetida (*Ferula asafoetida*)

No plant diseases have been recorded in *Ferula* species. This may be due to the presence of antifungal, antibacterial and antiviral properties in the various parts of the plants and natural resistance to diseases.

3

General Characters of Pathogens Affecting Spices

1. Genus - *Pythium* (Type species - *Pythium aphanidermatum*)

'Pythium' is a large genus comprising of more than 200 species. Species of *Pythium* are either terrestrial or aquatic. They are found world wide and most of them are plant pathogens on many economically important cultivated crops, as well as wild plants. They are facultative saprophytes and can live saprophytically in the soil and plant debris for indefinite periods in the absence of suitable hosts. Pre-emergence and post-emergence damping off of seedlings of numerous angiospermic plants, root rot, rhizome rot and fruit rot are the most common diseases caused by species of *Pythium*.

Pythium debaryanum, Pythium aphanidermatum, Pythium myriotylum, Pythium indicum, Pythium vexans and *Pythium graminicolum* are some of the most important plant pathogens, that cause damping off, seed decay, soft rots and root rots of many economically important crops, such as chillies, eggplant, tobacco, tomato, castor, potato, ginger etc.

Life cycle of *Pythium*

The mycelium of the fungi consists of slender, hyaline, coenocytic hyphae. The hyphae grow both inter- and intra-cellularly. No haustoria are produced.

In the asexual reproductory phase, the fungi produce globose to oval terminal or intercalary sporangia. During zoospore formation, a thin-walled, bubble-like vesicle is formed at the tip of a long tube produced from the sporangium. Inside the vesicle 8 - 20 zoospores are delimited. After maturity the vesicle bursts and the zoospores are liberated. The kidney-shaped zoospores having two lateral flagella attached to the concave region swim in water droplets, come to rest, shed their flagella and encyst as a globular thick-walled spore. The encysted zoospore germinates by producing a germ tube,penetrate into the host and form new mycelium inside the host. In some species lobulated sporangia are formed. From the lobe a vesicle is formed and zoospores are produced in the vesicle.

In the sexual reproductory phase the fungi produce thick-walled, smooth, globular oospore or zygote. Oospores are formed after fertilization of the egg in the oogonium (female reproductory organ) by a nucleus from the antheridium (male reproductory organ). The oospore germinates after a period of rest. At lower temperatures between 10°-17°C, the oospore produces motile zoospores from a vesicle formed at the end of a long germ tube, while at high temperature (28°C), the oospore germinates by a germ tube and penetrates into the host directly.

Species of *Pythium* causing diseases of spices-*Pythium aphanidermatum* on ginger, turmeric, garlic, coriander, fenugreek, cumin and black cumin; *P.myriotylum* on ginger; *P. vexans* on ginger & cardamom; *P. ultimum* on basil

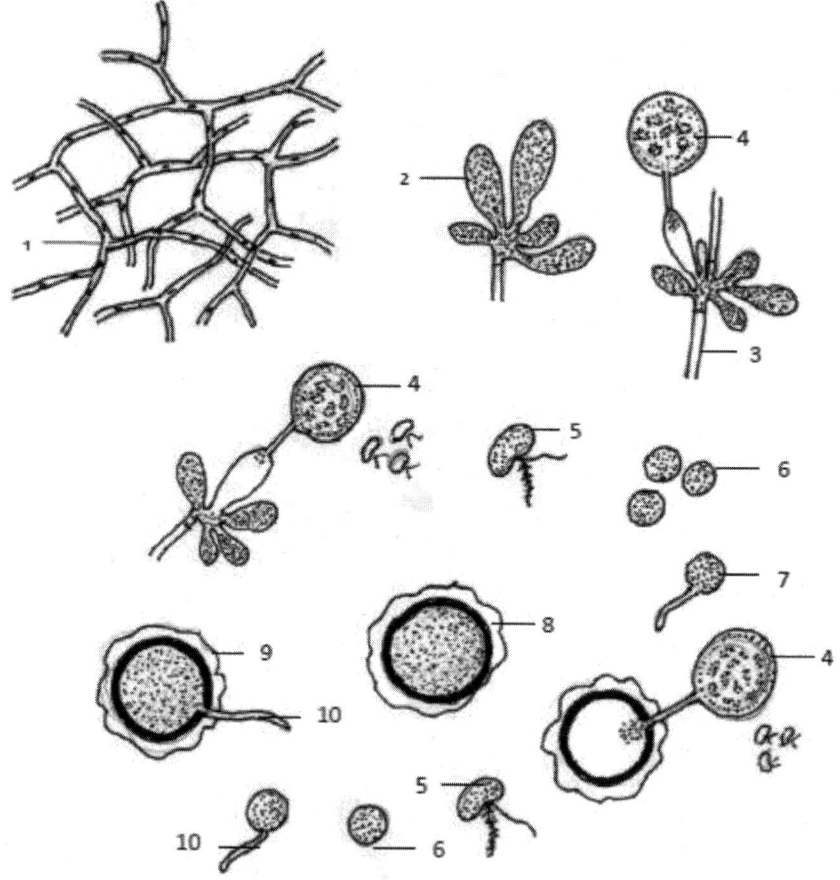

1. Coenocytic mycelium, 2. Terminal sporangium, 3. Intercalary sporangium, 4. Vesicle, 5. Zoospore, 6. Encysted zoospore, 7. Germinating encysted zoospore, 8. Oospore, 9. Germinating oospore, 10. Germ tube

Fig. 1. Type species - *Pythium aphanidermatum* (Soft rot of ginger)

2. Genus - *Phyllosticta* (Type species - *Phyllosticta zingiberii*)

Genus ***'Phyllosticta'*** comprises of many common and destructive plant pathogens. They typically infect thc foliage and cause tannish-gray leaf spots with dark brown to purple borders. Some species may also infect fruits and stems resulting in huge economic losses.

Leaf spot of ginger caused by *Phyllosticta zingiberii;* 'Freckle disease' caused by *P. musarum;* 'citrus black spot' caused by *P. citricarpa:* 'apple blotch' caused by *P. solitaria;* 'Frog's eye leafspot' of red gram caused by, *P. cajani;* leaf spot of ground nut caused by *P. arachidis;* leaf spot of tobacco caused by *P. nicotianae* are important diseases caused by different species of pathogens belonging to this genera.

Life cycle of *Phyllosticta*

The mycelium of fungi belonging to this genus is well-developed, septate, intra- and inter-cellular. The vegetative phase is followed by the reproductory phase in which the fungi produce asexual fruiting bodies called pycnidia. Pycnidia appear as minute, pin point-like, black dots on the upper surface of the diseased spots. Pycnidia are globose or flask-shaped, dark-colored, leathery and are sunken in the leaf tissue. They are provided with a narrow pore or ostiole. Pycnidiospores (conidia) are borne at the tips of very short, hyaline phialides arising from the hymenium situated at the bottom of the pycnidia. Conidia are produced in succession in large numbers. They are single-celled, hyaline, oblong with rounded ends. Ultimately the conidia are set free and come out through the ostiole. The conidia germinate by producing a germ tube and infect the host plant.

Species of *Phyllosticta* causing diseases in spices - *Phyllosticta zingiberii* on ginger; *P. elettariae* on cardamom; *P. murrayae* on curry leaf; *P. tamarindicola* on tamarind.

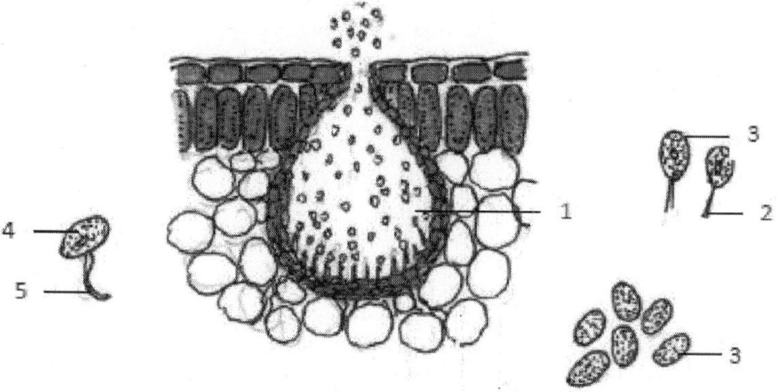

1. Pycnidia, 2. Phialide, 3. Pycnidiospore, 4. Germinating pycnidiospore, 5. Germ tube

Fig. 2. Type species - *Phyllosticta zingiberii* (Leaf spot of ginger)

3. Genus - *Colletotrichum* (Type species - *Colletotrichum zingiberis*)

'Colletotrichum' is an economically important plant pathogenic genus distributed worldwide. The genus is known to encompass about 190 species. Most of the species are endoparasites and cause diseases on a wide variety of cultivated and wild plants covering more than 30 plant genera. Most common disease called 'Anthracnose' is caused by *Colletotrichum gloeosporioides.*

Many species of *Colletotrichum* are seed-borne. However the pathogen can survive in the soil as a saprophyte on infected plant debris and continue to produce spores. The spores are disseminated by air currents and also by water splash. The ascospores released from the sexual organ (perithecium) can also initiate infection.

Anthracnose is primarily a leaf spot caused by some fungal pathogens belonging to *Colletotrichum* that appears initially as small, circular to irregular, yellow or brown, slightly sunken spots. The spots enlarge, turn dark and may cover the entire leaf area followed by necrosis under favorable conditions. Numerous pin point-like dots representing the acervuli are seen on the necrotic spots.

Anthracnose of banana fruits, mango fruits, leaf spot, blossom blight and wither tip of mango, anthracnose of citrus, leaf spot of pomegranate, anthracnose of cashew, leaf blight of jasmine, anthracnose or 'Pollu' disease of pepper, leaf spot of ginger, inflorescence die-back of arecanut etc., caused by *Colletotrichum gloeosporioides*; red rot disease of sugarcane caused by *C.falcatum*; anthracnose and die-back of tapioca caused by *C. manihotis*; ripe fruit rot and die-back of chillies, leaf spot of turmeric, rose etc., caused by *C. capsici*; leaf spot of ginger caused by *C. zingiberis*; 'Smudge' disease of onion caused by *C. circinans*; red spot disease of sorghum, maize etc. caused by *C. graminicola*; anthracnose of beans, lab lab and leaf spot of onion caused by *C. lindemuthianum;* anthracnose of cucurbits caused by *C. lagenarium;* anthracnose and fruit rot of eggplant and tomato caused by *C. phomoides;* anthracnose of coffee caused by *C. coffeanum*; anthracnose of pea caused *by C. pisi;* anthracnose of papaya caused by *C. papayae;* anthracnose of eggplant caused by *C. melongenae* are very serious and destructive diseases caused by different species of the genus *Colletotrichum.*

Life cycle of *Colletotrichum*

When a conidium (Conidiospore) dispersed either by wind or water splash lands on a susceptible host, it germinates by producing a germ tube with an appressorium at the end, which fastens the spore to the host. From the bottom of the appressorium a minute infection peg (infection hypha) emerges and enters the host by penetrating the cuticle and epidermal cell. Once established inside the host, the hypha ramifies by producing several branched, inter- and intra-cellular hyphae and develop into a mycelium. The hyphae absorb nutrients from the host cells and the cells become necrotic and develop typical symptoms. The mycelium thus formed by the branching hyphae is mostly localized around the point of infection. In the fructification phase, which follows the vegetative phase, the hyphae aggregate beneath the host epidermis in an entangled mass forming a saucer-shaped stroma called 'acervulus'. From the stroma numerous closely packed, short, erect, hyaline conidiophores arise like palisade cells. Conidia are produced from the tips of the conidiophores in succession. From the stroma dark-colored, long, bristle-like, septate, sterile structures called 'setae' are also formed. The conidia are embedded in a viscous fluid that swells under moist conditions. Continuous production of conidia exerts pressure on the host epidermis, which is pushed up and ruptured and the conidia are exposed as a pinkish, slimy mass at the opening of the acervulus. The setae also exert pressure and aid in rupturing the host epidermis.

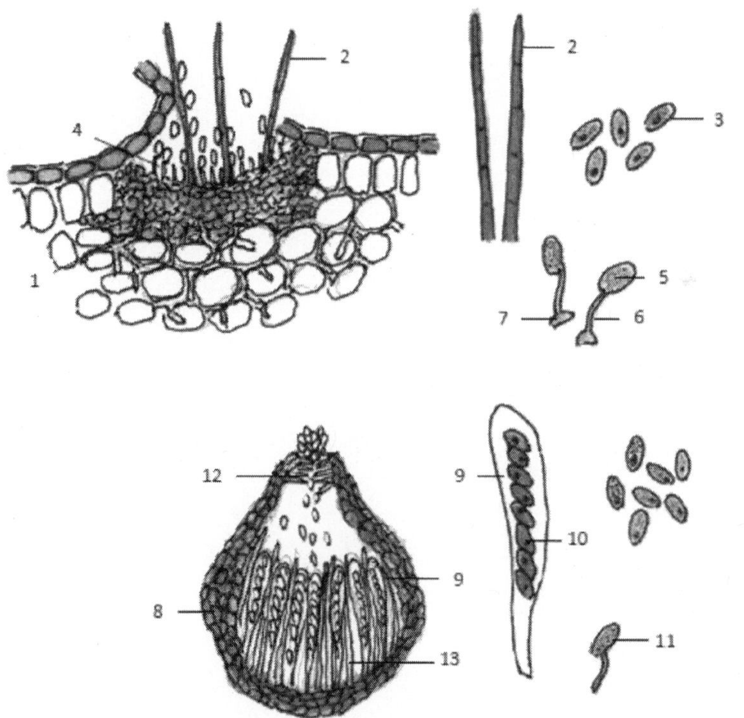

1. Acervulus, 2. Seta, 3. Conidium, 4. Conidiophore, 5. Germinating conidium, 6. Germ tube, 7. Appresorium, 8. Perithecium, 9. Ascus, 10. Ascospore, 11. Germinating Ascospore, 12. Periphyses, 13. Paraphyses

Fig. 3. Type species - *Colletotrichum gloeosporioides* (Anthracnose of pepper)

The perfect or aascigerous stage, *Glomerella* is rarely found in nature. Whenever the perfect stage occurs, the fruiting body or perithecia are found in the dead plant tissues and in the perfect stage the fungus is not pathogenic. The perithecia are dark colored, and sub globose to globose in shape with a rather prominent short neck with an opening called ostiole. Periphyses, which are short sterile appendages line inside the ostiolar opening. The hymenium, which consists of an assemblage of asci and paraphyses is formed inside the bottom of the perithecium. The asci are club-shaped and contain 8 ascospores arranged serially. Sterile long paraphyses appear in-between the asci. On maturity, the asci break open and the ascospores are liberated. The ascospores are then pushed out through the ostiolar opening The ascospores are also capable of causing disease

Species of *Colletotrichum* causing diseases of spices - *Colletotrichumg-loeosporiodes* on pepper, cardamom, clove, garlic, coriander, cinnamon, cassia, nutmeg, vanilla and curry leaf ; *C. zingiberis* on ginger; *C. necator* on pepper; *C. capsici* on chillies, turmeric and coriander; *C. vanillae* on vanilla; *C. graminicola* on lemon grass

4. Genus - *Fusarium* (Type species - *Fusarium oxysporum* f.sp. *corianderii*)

The Genus '***Fusarium***' contains over 20 species complexes. They are distributed throughout the world and some of them cause very serious wilt disease in several crops that kills the entire plant. The filamentous fungi are facultative parasites and live in the organic matter present in the soil. They produce three kinds of asexual spores viz., macroconidia, microconidia and chlamydospores.

Wilt of banana caused by *Fusarium oxysporum* f.sp. *cubense;* wilt of tomato caused by *F. oxysporum* f.sp. *lycopersici*; wilt of cotton caused by *F. oxysporum* f.sp. *vasinfectum;* wilt of gingelly caused by *F. oxysporum* f.sp.*sesami*; wilt of cucurbits caused by *F. oxysporum* f.sp. *melonis*; wilt of pea caused by *F.oxysporum* f.sp. *pisi*; foot rot of rice, 'Pokkah boeng' or twisted top disease of sugarcane, fruit rot of banana and citrus, sett rot or stem rot of sugarcane etc. caused by *F. moniliforme;* wilt of red gram caused by *F. udum* are very important diseases caused by species of this genus.

Life cycle of *Fusarium*

The pathogen enters the host plants through the root system. The mycelium produced by the fungi is septate, much branched, inter- and intra-cellular. After entry into the host, the mycelium reaches the vascular region, grows and multiplies in the xylem vessels. The mycelium, as well as the micro conidia produced in large numbers plug the vessels and obstruct the movement of water and mineral nutrients to the above ground parts. Further, the pathogen excretes certain toxic substances in the plant tissues. The toxin, 'fusaric acid' produced by the pathogen in the vascular system is translocated throughout the plant and this toxin is primarily responsible for the wilting. The pathogen is also known to produce certain 'pectinolytic', 'cellulolytic' and 'proteolytic enzymes', which cause disintegration of the cell walls. These enzymes, which produce a gummy substance in the vascular vessels is also partly responsible for plugging the vessels.

In advanced stages of infection, the pathogen forms whitish to grayish or bluish weft of mycelium on the collar region near the ground level and forms stroma. The pathogen produces both macro- and micro conidia. Macroconidia are formed on sporodochia arising from the stroma. Sometimes conidia are formed directly from the hyphae. Macroconidia are hyaline sickle-shaped or crescent-shaped, septate, pointed at both ends and have a pedicel. The macroconidia appear light buff to reddish-orange in mass. Microconidia are hyaline, oval-shaped, non-septate or rarely single-septate. Terminal or intercalary, thick-walled, spherical, single-celled chlamydospores are also produced. Sometimes any one cell of a macroconidium is converted into a chlamydospore.

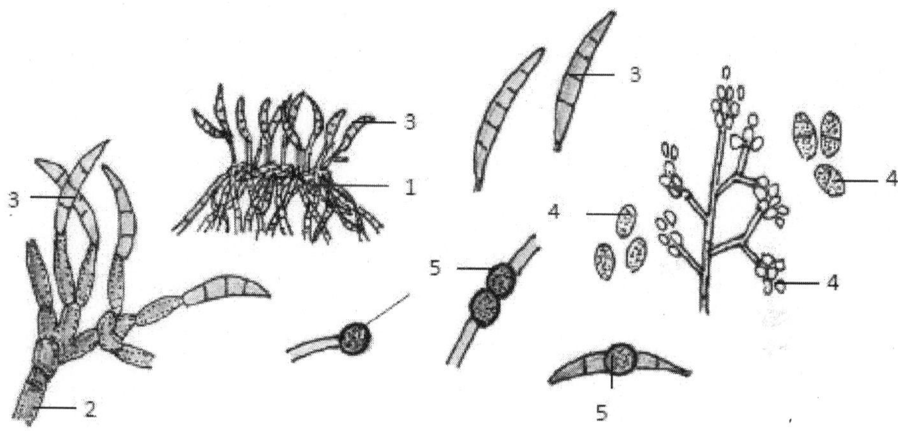

1. Sporodochium, 2. Conidiophore, 3. Macroconidia, 4. Microconidia, 5. Chlamydospore

Fig. 4. Type species -*Fusarium oxysporum*f.sp. *corianderii* (*Fusarium* wilt of coriander)

Species of *Fusarium* causing diseases in spices - *Fusarium oxysporum* on cardamom and fenugreek; *F. oxysporum* f.sp. *zingiberis* on ginger; *F. oxysporum* f.sp. *cepae* and *F. culmorum* on garlic; *F. oxysporum* f.sp. *corianderii* on coriander; *F. oxysporum* f.sp. *cumini* on cumin and black cumin; *F. oxysporum* f.sp. *vanillae* and *F. batatis* on vanilla; *F. oxysporum* f.sp. *basilicum* on basil; *Fusarium* sp. on clove, garlic and coriander; *F. equiseti* on fennel; *F. solani* on saffron; *F. oxysporum* f.sp. *cichorii* on chicory; *F. oxysporum* f.sp. *capsici* on chilli

5. Genus - *Rhizoctonia* (Type species - *Rhizoctonia solani*)

'Rhizoctonia' *(Moniliopsis)* is a genus of anamorphic fungi in the Family - Ceratobasidiaceae. They do not produce any kind of asexual or sexual spores but are composed of hyphae and sclerotia (hyphal propagules). Species of *Rhizoctonia* are facultative parasites and cause diseases in several commercially important crops. Now the genus is restricted to the type species *Rhizoctonia solani* and its synonyms. It is included in the Genus - *Thanatephorus*.

Pre-emergence and post-emergence damping off of seedlings of egg plant, crucifers, cotton etc., sheath blight of rice, root rot of tomato, pea, sugar beet etc., root and stem rot of cluster beans, crown rot of carrot, black scurf and stem canker of potato, foot rot of chrysanthemum, collar rot of coffee, etc. caused by *Rhizoctonia solani*; charcoal rot or stem rot of gingelly, sugar beet, cotton etc. caused by *Rhizoctonia bataticola* are very important diseases caused by species of *Rhizoctonia*. *Rhizoctonia solani* also causes hypocotyl and stem cankers on mature plants of tomatoes, potatoes and cabbages. Strands of mycelium and sometimes sclerotia appear on the affected area. Roots turn brown and die after a period of time. These are some of the most important and destructive diseases caused by species of the genus.

Life cycle of *Rhizoctonia*

Rhizoctonia solani can survive in the soil for many years in the form of sclerotia. The sclerotia consist of a mass of loosely interwoven hyphae and are provided with thick outer layers for survival. The mycelium can also remain alive in the soil for long periods of time as a saprophyte. The mycelium is much branched and the hyphal cells are multinucleate. The hyphae tend to branch at right angles, which is a characteristic feature of *Rhizoctonia*. A septum near each hyphal branch and a slight constriction at the branch are also characteristic. The fungus is attracted to the host plants by chemical stimuli released by the growing host plant, as well as decomposing plant residue. The pathogen can gain entry into the host by direct penetration of the cuticle and epidermis or through natural openings or through injuries. After successful invasion and establishment the pathogen causes necrosis and sclerotial formation around the infected tissue. The enzymes released by the pathogen break down adjacent cell walls and the pathogen spreads and occupies new areas of the host.

The teleomorph stage of *Rhizoctonia solani* is *Thanatephorus cucumeris*. It forms club-shaped basidia with four apical sterigmata on which oval, hyaline, single-celled basidiospores are borne.

Species of *Rhizoctonia* causing diseases in spices-*Rhizoctonia solani* on ginger, pepper, cardamom, turmeric, garlic, coriander, fenugreek and chicory; *R. bataticola* on pepper,; *R. syzygii* on clove; *Rhizoctonia* spp. on cardamom and turmeric.

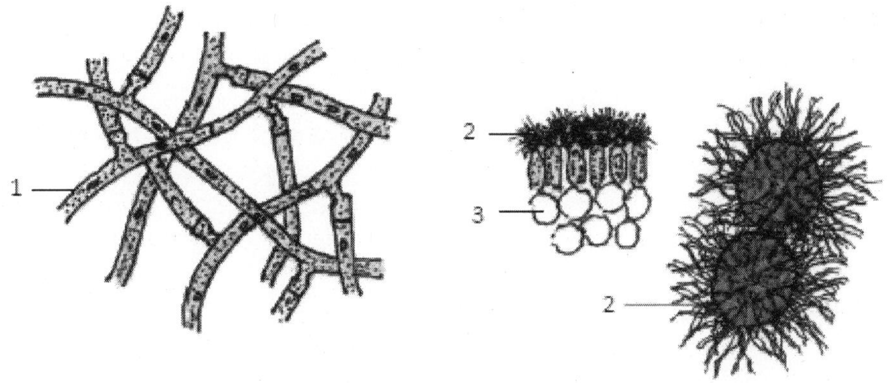

1. Mycelium, 2. Sclerotium, 3. Host plant

Fig. 5. Type species - *Rhizoctonia solani* (Leaf rot and blight of pepper)

6. Genus - *Sclerotium* (Type species - *Sclerotium rolfsii*)

Members of genus *Sclerotium* produce only sclerotia and no other fruiting bodies or spores. The genus includes more than 40 plant pathogenic species. *Sclerotium rolfsii* is a type species of this genus. *Sclerotium rolfsii* causes diseases on hundreds of plant species including field, vegetable, fruit and ornamental crops. All plant parts, such as roots, stems, leaves and fruits may be affected Infection usually is restricted to plant parts in contact with the soil.

Root rot of tomato, chillies, tapioca, sugar beet etc., seedling blight of apple, damping off of egg plant, rot of potato, basal wilt of pepper, stem and pod rot of groundnut, stem and root rot of tobacco, wilt of betel vine etc. caused by *Sclerotium rolfsii*; stem blight of bean caused by *S. bataticola;* white rot of onion and garlic caused by *S. cepivorum*; stem rot of rice caused by *S. oryzae.* are some of the serious diseases caused by different species of the genus.

Life cycle of *Sclerotium*

Members of this genus produce only sclerotia and no other fruiting bodies or spores. The genus includes more than 40 plant pathogenic species. The fungi produce abundant white mycelium on the infected plants. Signs of infection include development of coarse, white, silky strands of mycelium that grows in a fan-shaped pattern on the soil surface, leaf litter or on the lower stem region. The hyphae are septate and much branched. After about 7 - 14 days, tan to brown, mustard seed-sized sclerotia are formed on the mycelial mat. Sclerotium is a vegetative resting body composed of a compact mass of hardened fungal mycelium containing food reserves. The sclerotia remain dormant until a host plant's root exudates stimulate germination. Sclerotia undergo either hyphal or eruptive germination. In hyphal germination individual strands of hyphae grow from the sclerotial surface, while in eruptive germination, aggregates of mycelium burst out through the sclerotial surface. Infection usually is restricted to plant parts in contact with the soil. The fungus kills tissues of affected plants by production of oxalic acid as well as cell-wall degrading enzymes.

Species of *Sclerotium* causing diseases in spices - *Sclerotium rolfsii* on ginger, pepper, garlic, fenugreek and saffron.

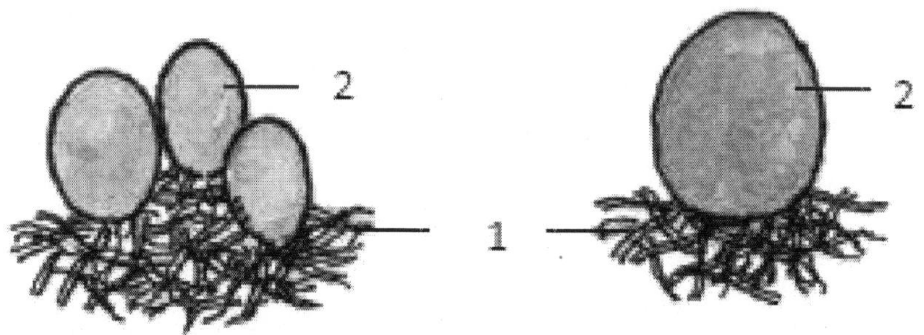

1. Mycelium, 2. Sclerotium

Fig. 6. Type species – *Sclerotium rolfsii* (Basal wilt of pepper)

7. Genus - *Phytophthora* (Type species - *Phytophthora capsici*)

'Phytophthora' is a large genus comprising of about 170 described species. They are distributed world wide and most of them are pathogens on dicotyledons and many species are plant pathogens of considerable economic importance. They cause root, crown and collar rots, wilts, leaf and stem blights, fruit and tuber rots, cankers and die-back in several economically important plants. They also cause many destructive diseases in forest trees. They are actually facultative parasites and can over winter in soil and infected plant debris as thick-walled resting spores, such as chlamydospores and oospores or as vegetative mycelium.

Late blight of potato caused by *Phytophthora infestans;* 'Black shank' of tobacco, leaf and root rot of betel vine, leaf fall, canker and pod rot of rubber, seedling blight of castor, cotton etc. caused by *P. parasitica;* bud rot of coconut and 'Koleroga' disease of areca nut caused by *P. palmivora*; root rot of citrus plants caused by *P. citrophthora;* quick wilt and slow decline of pepper caused by *P. capsici*; capsule rot of cardamom caused by *P. nicotianae*; leaf blight of cardamom caused by *P. meadii;* 'Stripe canker' of cinnamon caused by *P. cinnamomi* are some of the very important diseases caused by different species of *Phytophthora.*

Life cycle of *Phytophthora*

The mycelium of fungi belonging to this genus is slender, profusely branching, coenocytic, multinucleate, mostly intercellular and sometimes intracellular. No haustoria are produced. In the asexual reproductory phase the fungi produce sporangia and chlamydospores. Sporangia are produced singly from the tips of short hyphal stalks called sporangiophores or from the tips of specialized, sympodially branched sporangiophores arising from the internal mycelium as in the case of *Phytophthora infestans..*The sporangia are pear-shaped and have prominent papilla. Zoospores are formed within the sporangium and after maturity they come out by breaking open the papilla. In some cases, a vesicle is formed by the swelling of the papilla and the zoospores produced within the sporangium move into the vesicle. The wall of the vesicle ruptures and the zoospores are liberated.

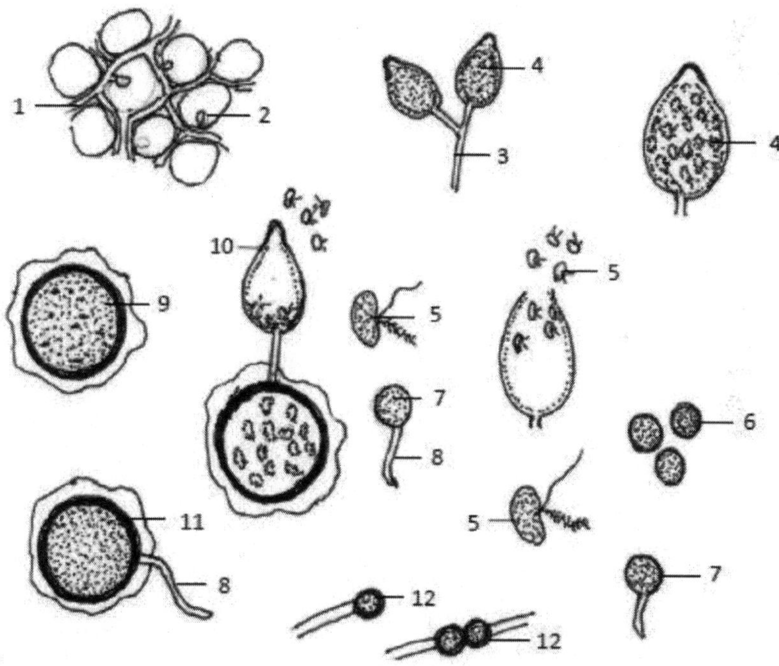

1. Intercellular mycelium, 2. Haustorium, 3. Sporangiophore, 4. Sporangium, 5. Zoospore, 6. Encysted Zoospore, 7. Germinating zoospore, 8. Germ tube, 9. Oospore, 10. Vesicle, 11. Germinating oospore, 12. Chlamydospore

Fig. 7. Type species - *Phytophthora capsici* (Foot rot of pepper)

The zoospores are bean-shaped and are motile by means of two flagella arising laterally from the concave portion. They swim about in free water droplets for a short time, come to rest and encyst. The encysted zoospores are spherical and thick-walled. After a period of short rest they germinate by means of a germ tube and penetrate into the host. Asexually the fungi also produce thick-walled, terminal or intercalary resting spores or chlamydospores.

In the case of sexual reproductory phase, the fungi produce oospores as a result of fertilization of the oogonium (female organ) by an antheridium (male organ). They are spherical, smooth and thick-walled resting spores. They germinate by means of a germ tube and infect the host.

Species of *Phytophthora* causing diseases in spices - *Phytophthora cinnamomi* on cinnamon and cassia; *P. Capsici* on pepper and chilli pepper; *P. nicotianae* on cardamom; *P. meadii* on Cardamom and Vanilla

8. Genus - *Cercospora* (Type species - *Cercospora ricinella)*

The genus ***'Cercospora'*** is a large one and comprises of about 3000 described species. It represents a group of important plant pathogenic fungi with a wide geographic distribution. The fungi belonging to this genus are commonly associated with leaf spots on a broad range of economically important crops. Most of the species have their perfect stage in the genus *Mycosphaerella.*

'Tikka' leaf spot of groundnut caused by *Cercospora personata* and *C. arachidicola;* narrow brown leaf spot of rice caused by *C. oryzae;* 'Frog's eye' leaf spot of tobacco caused by *C. viticola;* brown leaf spot of grapevine caused by *C. viticola*; leaf spot of egg plant caused by *C. melongenae* and *C. deightonii;* leaf spot of chillies caused by *C. capsici;* leaf spot of tapioca caused by *C. manihotae*; leaf spot of castor caused by *C. ricinella*; 'Sigatoka'disease of banana caused by *C. musae* are very important diseases caused by species of this genus.

Life cycle of *Cercospora*

The conidia of the pathogen landing on the host germinate and enter the host either through the stomata or by direct penetration of the cuticle and epidermal cells. The hypha that enters the host develops into well branched mycelium mostly around the point of infection. The mycelium is light brown in color, septate, inter- and intracellular. Before fructifications are formed, the hyphae collect beneath the epidermis in a stroma and from the stroma a bunch of about 10 conidiophores emerge either through the stomata or directly by penetrating the host epidermis. The conidiophores are dark colored and are thicker than the ordinary hypha, unbranched, non-septate or 1- 2 septate. Conidia are borne at the tips of the conidiophores singly and are detachable easily. Conidia are sub-hyaline, thin-walled, long, obclavate, the basal portion slightly broader than the apex and 6 - 7 septate.. Meanwhile the conidiophore continues to grow further forming a knee bend and produce another conidium at its tip. The conidiophores are geniculate and a scar is discernible at each of the knee bends from where the conidia had dislodged.

Species of *Cercospora* causing diseases in spices - *Cercospora zingiberi* on cardamom; *C. traversiana* on fenugreek; *C. ocimicola* on basil; *C. capsici* on chilli pepper

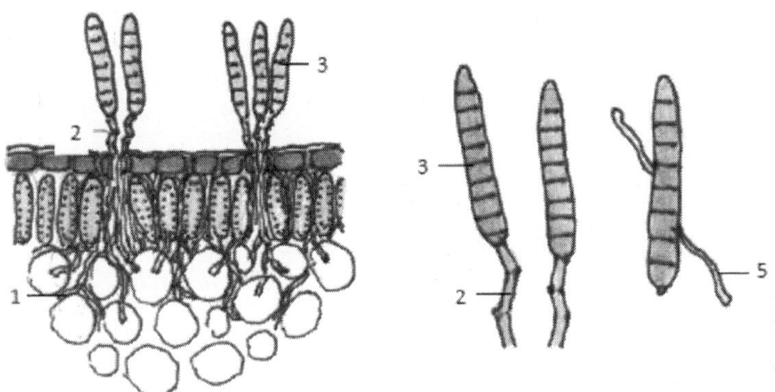

1. Mycelium, 2. Conidiophore, 3. Conidium, 4. Germinating conidium, 5. Germ tube

Fig. 8. Type species - *Cercospora traversiana* (Leaf spot of fenugreek)

9. Genus - *Cylindrocladium* (Type species - *Cylindrocladium quinquiseptatum*)

Genus ***'Cylindrocladium'*** comprises of about 52 species distributed worldwide. Most of them are parasites on plants. They attack over 100 woody ornamental shrubs and trees, as well as foliage plants. They cause diseases, such as damping off, wilts, root rots, stem cankers, crown rots and leaf spots.

Important diseases such as leaf rot of clove, leaf blight of eucalyptus, leaf rot/ brown rot of vanilla, leaf spot of rubber etc. are caused by *Cylindrocladium quinquiseptatum.* Seedlings of many species of conifer and hard wood trees, such as pines, wall nuts, poplars, oaks and several ornamental shrubs are affected by *Cylindrocladium scoparium, C. floridanum, C. crotalariae* etc.

The fungus is primarily soil-borne. The fungus over winters as microsclerotia, which can survive for several years in soil or on infected plant debris. Germination of microsclerotia is triggered when they come into contact with root exudates of host plants. Once germination is initiated, penetration into the host roots occur within 24 hours. Septate hyphae colonize the roots and begin to produce masses of mycelium that produce microsclerotia within a few days.

Conidiophores are produced on the external plant tissue and form conidia that are dispersed by wind and water splash. Conidiophores are formed scattered over the leaf surface. The main axis of the conidiophore extends beyond the sporogenous zone to form a long sterile appendage known as a stipe extension. The stipe extension becomes narrower just below the swollen apical vesicle, which is hyaline and ellipsoidal in shape. The sporogenous portion of the conidiophore is composed of two or more bifurnicate lateral branches. The primary branches give rise to secondary and sometimes tertiary branches with progressively smaller cells, each branch ending in two or more phialides. Conidia are produced from the phialides. Conidia are straight, cylindrical and one septate. Chlamydospores develop in abundance either in chains or clumps and appear blackish. The chlamydospores are referred to as microsclerotia.

Life cycle of *Cylindrocladium*

Species of *Cylindrocladium* causing diseases in spices - *Cylindrocladium quinquiseptatum* on Clove and Vanilla

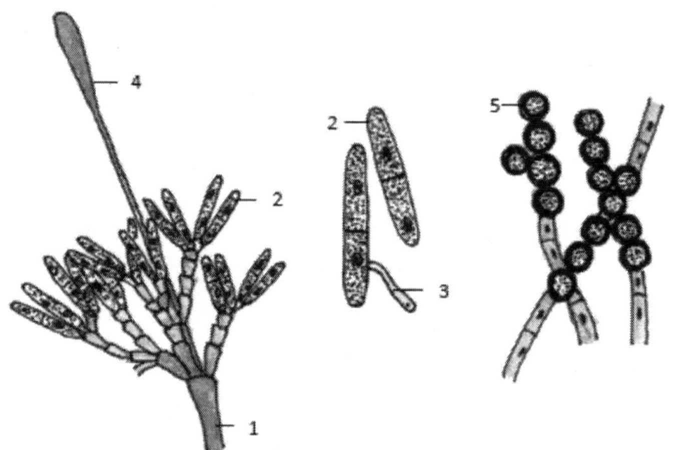

1. Conidiophore, 2. Conidium, 3. Germinating conidium, 4. Stipe extension, 5. Chalmvdospore

Fig. 9. Type species - *Cylindrocladium quinquiseptatum* (Leaf rot ofclove)

10. Genus - *Cryphonectria* (Type species - *Cryphonectria cubensis*)

Species of the genus ***'Cryphonectria'*** include some of the world's most important and serious tree pathogens and attack a wide variety of woody hosts. They cause the deadly canker disease and die-back.

'Chestnut blight' caused by *Cryphonectria parasitica;* canker and die-back of clove and Eucalyptus caused by *C. cubensis, C. havanensis* and *C. eucalypti;* canker and die-back of Coccoloba caused by *C. coccolobae* are some of the worst tree diseases caused by species belonging to the genus.

Life cycle of *Cryphonectria*

The spores either conidia or ascospores getting access to wounds in the trees germinate and enter the host. The hyphae multiply rapidly in the bark tissue and produce spore bodies. The ascostromata produced are partly immersed in the bark tissue. In some cases erumpent, orange-colored ascostromatic tissue are formed. Ascostromata are black in color and are globose. They have black, cylindrical, protruding neck. Periphyses are present. In each stroma 1-9 perithecia may be found. The perithecia have no distinct walls and are locular. The asci formed inside the perithecial locules are eight-spored, biseriate and ellipsoid. Ascospores are hyaline, one-septate, fusoid to oval with tapered apices.

Conidiomata occur on the top of ascostromata. Conidiomata are multilocular and each locule has a long neck. A single locule may be connected to one or several necks. Conidiophores are hyaline with a cell, either single or branched. Conidiogenous cells are phialidic. Conidia are hyaline, non-septate, oblong and are extruded through the neck as bright spore tendrils or droplets.

Species of *Cryphonectria* causing diseases in spices - *Cryphonectria cubensis* on Clove

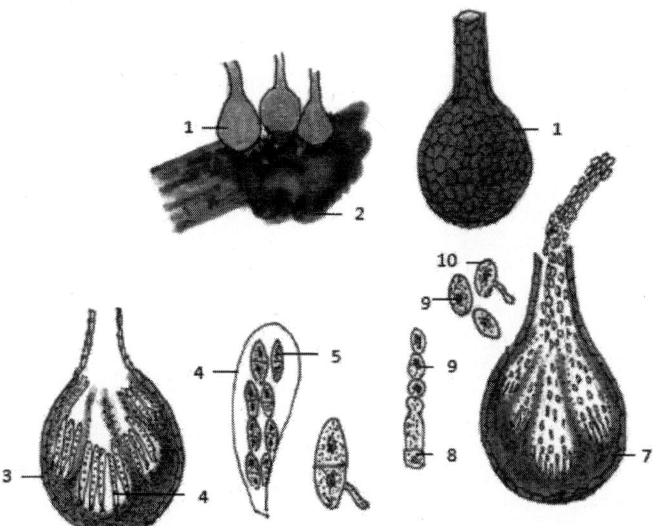

1. Ascostromata, 2. Bark tissue, 3. Perithecium, 4. Ascus, 5. Ascospore 6. Germinating ascospore, 7. Conidiomata, 8. Conidiogenous cells, 9. Conidia, 10. Germinating conidium

Fig. 10. Type species - *Cryphonectria cubensis* (Die-back / canker of clove)

11. Genus - *Valsa* (Type species - *Valsa eugeniae*)

Genus ***'Valsa'*** comprises of about 70 species distributed worldwide. Canker caused by species of this genus is one of the most serious and wide spread disease of plants including fruit trees, hard wood trees, shade trees and shrubs. They cause extensive damage and even death of trees by forming cankers on the bark and girdling branches or the main stem. The ascospores and conidia produced in large numbers are washed down to the soil and infect the absorbing and fibrous roots. When the roots are thus affected, they are unable to absorb water from the soil, resulting in sudden death of the affected trees.

Valsa eugeniae causes die-back and canker of mature clove trees as well as many other trees and shrubs. Canker of apple trees caused by *Valsa mali;* canker of almond caused by *V. pyri* are some of the serious diseases caused by species of this genus.

Life cycle of *Valsa*

The fungus is a typical wound parasite. Once it gains entry into the host, the mycelium spreads and travels down till it reaches a junction where the stem branches and develops cankers. Ascostromata are found immersed in the bark of affected trees with black perithecia in locules of 4-12. Perithecia lack definite perithecial walls. Perithecia are globose with long osteoles. Asci are sessile, unitunicate, sub-clavate with rounded apex and are eight-spored. Ascospores are hyaline and aseptate. The conidiomata, which are also found embedded in the bark tissue are dark, mostly unilocular, globose or ovoid. The pycnidia in the conidiomata has a single common opening through which large number of conidia are exuded in a gelatinous matrix. Conidiophores are hyaline, branched and phialidic. Conidia are hyaline, allantoid and aseptate.

Species of *Valsa* causing diseases on spices - *Valsa eugeniae* on clove

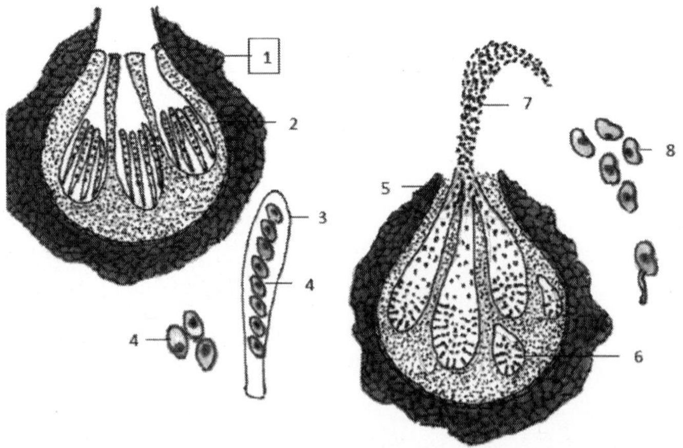

1. Ascostroma, 2. Perithecia in locules, 3. Ascus, 4. Ascospores, 5. Conidiomata, 6. Pycnidia, 7. Conidia extended in a gelatinous matrix, 8. Conidia, 9. Germinating conidium

Fig. 11. Type species - *Valsa eugeniae* (Sudden death of clove)

12. Genus - *Taphrina* (Type species - *Taphrina maculans*)

Genus '***Taphrina***' contains about 95 recognized species, which are all important pathogens of vascular plants. They induce hypertrophic malformations on various plant parts, such as buds, leaves, twigs, flowers and fruits producing diseases, such as leaf curl, leaf spot, leaf blotch, 'witches broom' etc.

Leaf curl of peach and almond caused by *Taphrina deformans;* 'Witches broom' of cherry caused by *T. cerasi;* curling and puckering of oak leaves caused by *T. coerulescens;* leaf spot of turmeric caused by *T. maculans* are important diseases caused by different species of the genus.

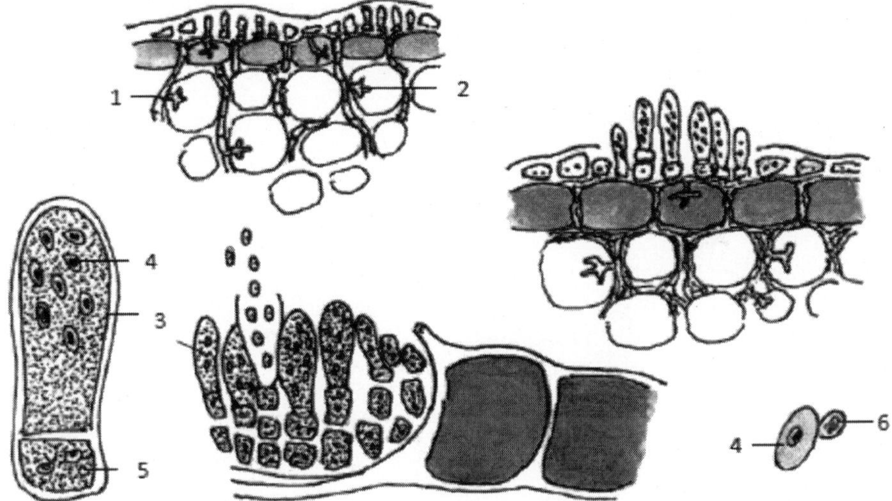

1. Mycelium, 2. Haustoria, 3. Ascus, 4. Ascospores, 5. Basak stalk cell, 6. Blastospore, 7. Blastospore budding from ascospore

Fig. 12. Type species - *Taphrina maculans* (Leaf spot of turmeric)

Life cycle of *Taphrina*

The mycelium is composed of septate hyphae consisting of binucleate cells or multinucleate cells with the nuclei occurring in pairs. The hyphae may be intercellular or sub cuticular or may grow within the walls of the epidermal cells. Asexual reproduction is by uninuleate, haploid, blastospores (conidia)) that bud from the ascospores. The blastospores may either bud and produce more blastospores or may germinate and infect the host. Karyogamy occurs between sister nuclei. The dikaryotic condition arises at the time of germination through a mitotic division of the nucleus of the ascospores or blastospores. As the hypha grows, conjugate division of the nuclei perpetuates the binucleate condition in the hyphal cells. Hyphae eventually aggregate in the sub cuticular region and break up into binucleate cells, which are the ascogenous cells. Karyogamy occurs within each ascogenous cell, which begins to elongate. At this stage the diploid nucleus divides mitotically. One daughter nucleus remains near the base of the cell, while the other moves toward the growing tip. Then a septum is formed between the two nuclei separating the cell into a basal stalk cell and an upper ascus mother cell, which is converted into an ascus. Meiotic and a subsequent mitotic division follows resulting in the formation of eight nuclei. Finally each nucleus develops into an ascospore. As the ascus mother cells enlarge, they exert pressure on the host cuticle and break through to form a compact surface layer of asci . No fruiting bodies are formed, as such the asci remain naked and release the ascospores into the air by bursting at the tip. After release the ascospores begin to bud and produce numerous blastospores. The blastospores eventually produce germ tubes and infect the host.

Species of *Taphrina* causing diseases on spices - *Taphrina maculans* on turmeric

13. Genus - *Peronospora* (Type species - *Peronospora parasitica*)

The genus ***'Peronospora'*** comprises of about 75 species. Diseases caused by species of *Peronospora* are generally referred to as 'downy mildew'. The fungi produce thin, grayish-white, scattered patches of fine, cottony, downy growth on the under surface of leaves and hence they are called as downy mildew. All species of this genus are obligate parasites.

Downy mildew of onion caused by *Peronospora destructor;* downy mildew (blue mould) of soybean caused by *P. manchurica;* downy mildew of tobacco caused by *P. tabacina;* downy mildew of mustard, cabbage, cauliflower and other cruciferous crops caused by *P. parasica* are quite destructive diseases caused by species of *Perospora.*

Life cycle of *Peronospora*

The mycelium is strictly intercellular. Large elongated or branched haustoria are produced from the hyphae, which penetrate the host cells and absorb nutrients from the host. Many haustoria may invade a single cell and fill the cell almost completely. At the fructification stage numerous, erect, branched sporangiophores emerge through the stomata on the under surface of leaves from the endophytic mycelium. The sporangiophores bifurcate 6 - 8 times at the top portion at acute angle with each other. The final branches, the sterigmata are slender, pointed and bear a single sporangium at the tip. The sporangia are broadly oval and hyaline. They germinate directly by a lateral germ tube and enter the host by penetrating the host epidermis or through the stomata.

Later, sexual reproduction takes place by copulation of an antheridium and an oogonium resulting in the formation of an oospore. The oospores are furnished with a thick wall with crest-like folds. They are globose and yellowish-brown in color. They germinate by producing a germ tube and enter the host. No zoospores are formed either by the sporangia or by the oospores.

Species of *Peronospora* causing diseases on spices - *Peronospora destructor* on garlic and onion; *P. trigonellae* on fenugreek; *P. umbellifarum* on aniseed; *P. parasitica* on mustard, black mustard and white mustard; *P. belbahrii* on basil

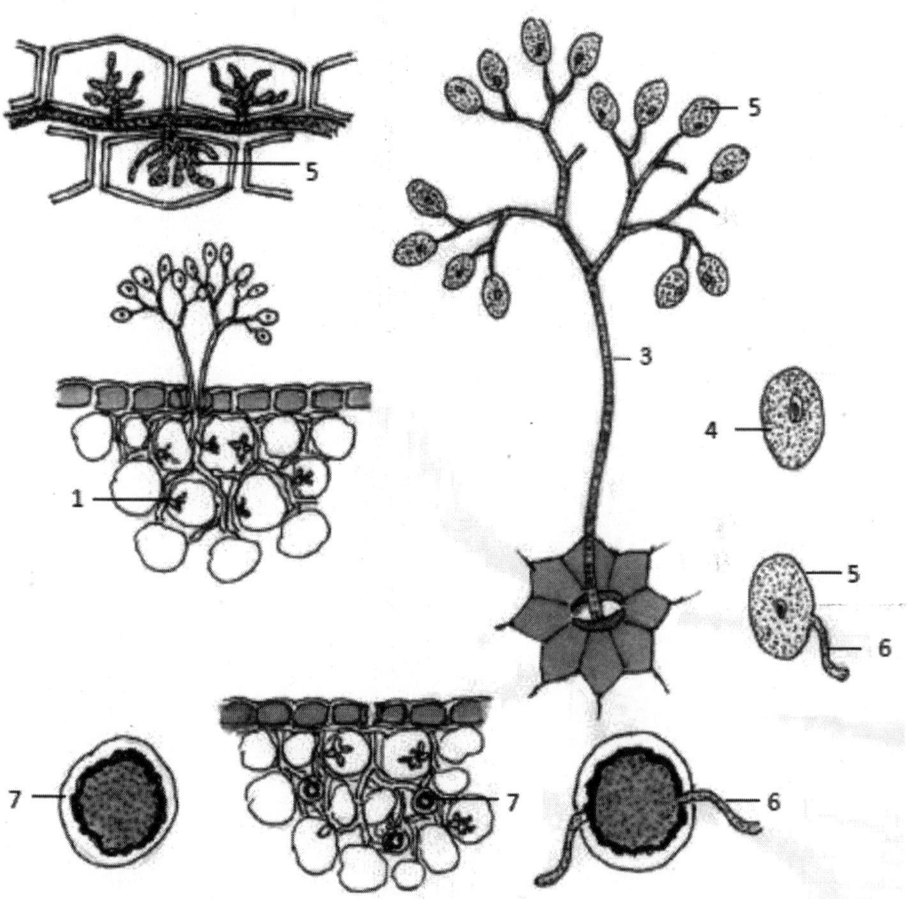

1. Intercellular mycelium, 2. Haustoria, 3. Sporangiophore 4. Sporangium, 5. Germinating sporangium, 6. Germ tube, 7. Oospore, 8. Grminating oospore

Fig. 13. Type species - *Peronospora parasitica* (Downy mildew of mustard)

14. Genus - *Leveillula* (Type species - *Leveillula taurica)*

There are 13 species of '***Leveillula***', which are parasitic on 50 host plant species. Of all the species *Leveillula taurica* is the most important one. All the species are obligate plant parasites. *Leveillula taurica* causes powdery mildew on many important crops, such as garlic, onion, cotton, cucumber, tomato, pepper, mulberry etc.

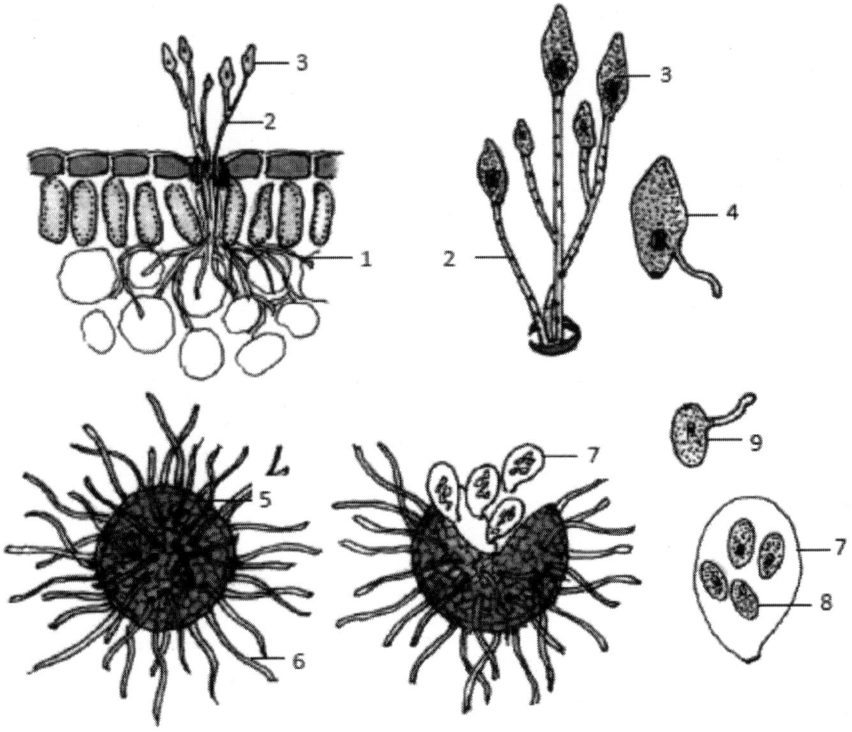

1. Mycelium, 2. Conidiophore, 2. Conidium, 4. Germinating conidium, 5. Cleistothecium, 6. Myceloid appendages, 7. Ascus, 8. Ascospore 9. Germinating ascospore

Fig. 14. Type species - *Leveillula taurica* (Powdery mildew of garlic)

Life cycle of *Leveillula*

The mycelium is endophytic, intra- and inter- cellular, branched and septate. The hyphae penetrate into the leaf through the stomata and spread between the mesophyll cells. A few days after infection and establishment of the pathogen inside the host, conidiophores are formed from the endophytic mycelium, which emerge out through the stomata and produce conidia from their tips on the surface of the leaves. The conidiophores are often branched. The conidia formed fall off as soon as they are formed and no conidial chains are formed. The conidia are lanceolate, hyaline and single-celled. Conidia do not require free water for germination and may germinate even at 0% humidity.

In the sexual reproductive phase, which follows the asexual reproductive phase, ascocarps called cleistothecia are formed in the infected crop residues above the soil surface. Cleistothecia, which are white in the beginning gradually changes to orange, reddish-brown and finally become black in color. They are globose, completely closed and have mycelioid appendages. Each cleistothecium contains 6-10 ovoid asci. Each ascus contains 3-5 ascospores, which are hyaline, elliptical and single-celled. Cleistothecia, which are the resting spores remain viable in the crop residue or soil for a long time. Under favorable climatic conditions the cleistothecial wall ruptures and the ascospores found inside the asci are released. The ascospores are dispersed by wind. When the spores find a suitable host, they germinate, enter the host through the stomata and cause fresh infection.

Species of *Leveillula* causing diseases on spices-*Leveillula taurica* on garlic, pepper, fenugreek and chilli pepper

15. Genus - *Alternaria* (Type species - *Alternaria brassicae*)

Genus ***'Alternaria'*** is a fairly large one comprising of about 50 species. A majority of the species are saprobic, while many of them are major plant pathogens and attack a wide variety of most economically important crops. The fungi are ubiquitous and many of them cause pre- and post-harvest damage to agricultural products including cereal grains, fruits and vegetables. In addition to spoiling a wide variety of foods, several *Alternaria* species produce secondary metabolites, such as phytotoxins, which are injurious to plants as well as mycotoxins, which are harmful to humans and animals.

Early blight of potato and tomato and fruit rot of chillies caused by *Alternaria solani;* fruit rot of apple, leaf spot of sugar beet etc. caused by *A, alternata;* rot of mango fruits caused by *A. tenuissima;* fruit rot of citrus caused by *A. citri;* leaf spot of crucifers caused by *A. brassicae* and *A. raphani;* leaf blight of jasmine caused by *A. jasmini Alternaria* blight of cotton caused by *A. macrospora;* blight of castor caused by *A. ricini* are some of the most serious diseases caused by species of this genus.

Life cycle of *Alternaria*

The mycelium of the fungus is septate, branched, light brown, which becomes darker with age. The hyphae are inter-cellular at first and later become intra-cellular. Conidiophores emerge in fascicles through the stomata from the necrotic zones of the spots. They are short, septate and dark colored. Conidia are produced in acropetal succession, either singly or in chains of up to four. Conidia also referred to as 'dictyospores' are rather large, multinucleate, rather long-beaked, straight or slightly curved, smooth, obclavate to muriform and dark colored, with 16 - 19 transverse septa and 0 - 8 longitudinal septa. They germinate in moist weather by putting forth germ tubes. From each conidium 5-19 germ tubes may arise and enter the host, either through the stomata or by penetrating the host epidermis.

Species of *Alternaria* causing diseases in spices - *Alternaria porri* on garlic and onion; *A. alternata* on coriander and aniseed; *A. burnsii* on cumin; *A. brassicae* and *A. brassisicola* on mustard, black mustard and white mustard

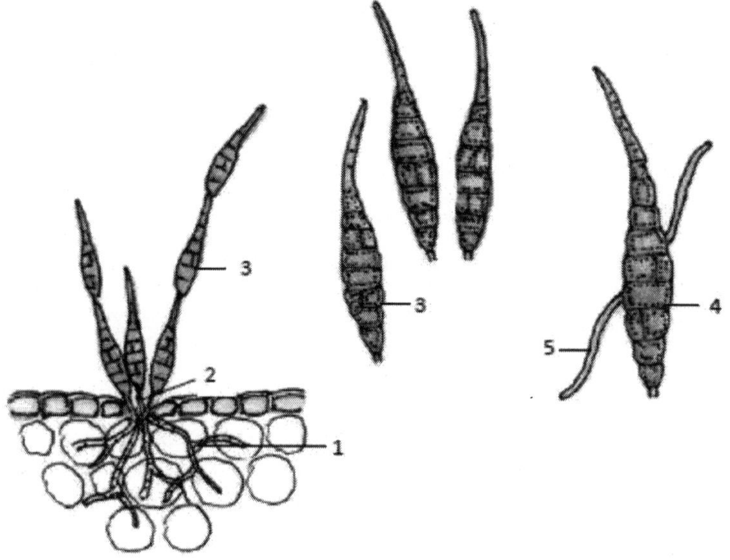

1. Mycelium 2. Conidiophore, 3. Conidia, 4. Germinating conidium, 5. Germ tube

Fig. 15. Type species - *Alternaria brassicae* (Leaf spot of crucifers)

16. Genus - *Aspergillus* (Type species - *Aspergillus niger*)

'Aspergillus' is a very large genus consisting of about 250 identified species of mould fungi. They are found in a variety of environments throughout the world. A vast majority of the species are saprophytes, while a few are pathogenic to humans, animals and plants. Most of the species reproduce asexually and produce conidia and are often referred to as 'conidial fungi'. Only a few species reproduce sexually and form cleistothecia containing asci and ascospores. *Aspergillus niger* group of fungi are commonly called as 'black Aspergilli' or 'black mould' because of the black-colored colony they produce. Many species of *Aspergillus* are usually found on exposed foodstuffs and cause decay. During this process some of them produce toxic substances called 'Mycotoxins', the most important one is 'aflatoxin', which is named after the species *Aspergillus flavus*. Mycotoxins cause several diseases in cattle. poultry and even in humans.

Aspergillus fumigatus, A. flavus, A. niger, A. terreus and many other spccics are parasitic on animals, birds and humans and cause a disease complex of the lungs known as 'Aspergillosis', that closely resembles 'tuberculosis'. Many species of *Aspergillus* grow on leather, paper and cloth fabrics and impart a musty, bad odor under humid weather conditions. Citric acid, gluconic acid, some enzyme substances and antibiotics are obtained from some species of *Aspergillus*.

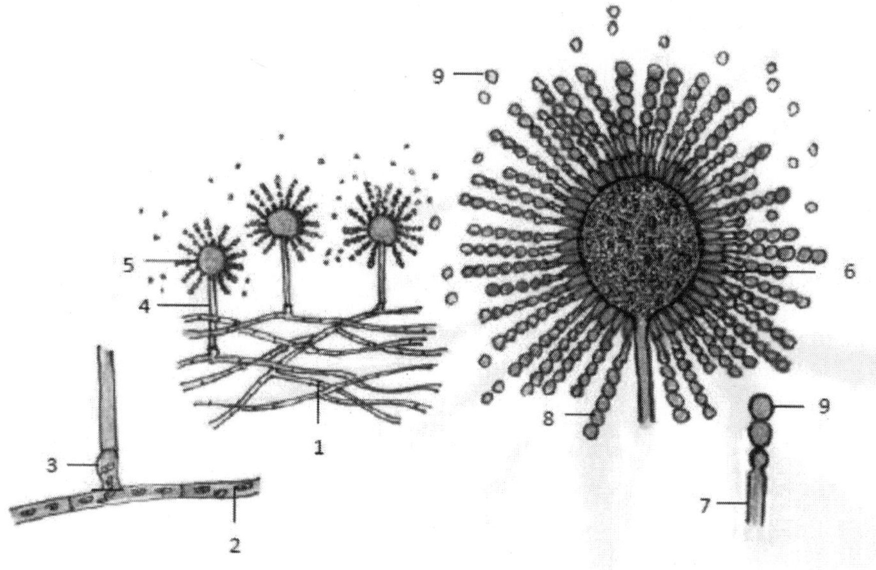

1. Mycelium 2. Hypha, 3. Foot cell, 4. Conidiophore, 5. Vesicle, 6. Conidiogenous cell, 7. Phialides, 8. Conidial chain, 9. Conidium

Fig. 16. Type species - *Aspergillus niger* (Black mould of garlic)

Life cycle of *Aspergillus*

The spores of *Aspergillus* are found abundantly floating in the air. The spore landing on a suitable substrate germinates and produces numerous hyphae that form the mycelium. The mycelium soon spreads across the surface of the substrate. The profusely branched hyphae develop into both vegetative and reproductive hyphae. The vegetative hyphae serve to absorb nutrients from the substrate, while the reproductive hyphae develop further to produce spores. The hyphae are septate, hyaline and the cells are multinucleate. The hyphal cell that branches to give rise to the conidiophore is called the 'foot cell'. The conidiophores are long, erect and terminate in a bulbous head called the 'vesicle'.

From the multinucleate vesicle, a large number of conidiogenous cells (mother cells) are produced over the entire surface of the vesicle. One or two layers of such cells (sterigmata) may be produced according to the species. The conidia bearing cells are typical phialides. The phialides cut off daughter cells in succession from their tips, one below the other in chains, which become the conidia. Conidia are globose, unicellular and rough-walled.

Species of *Aspergillus* causing diseases on spices - *Aspergillus niger* on garlic

17. Genus - *Stemphilium* (Type species - *Stemphilium vesicarium*)

The genus ***'Stemphylium'*** comprises of about 33 recognized species. Many species are saprophytic in nature, while others cause diseases on a number of important crops.

Leaf blight of onion and garlic caused by *Stemphylium vesicarium;* grey leaf spot of tomato and leaf blight of cotton caused by *S. solani;* leaf spot of alfalfa caused by *S. botryosum;* black rot of carrot caused by *S. radicinum* etc. are important disease caused by species of this genus.

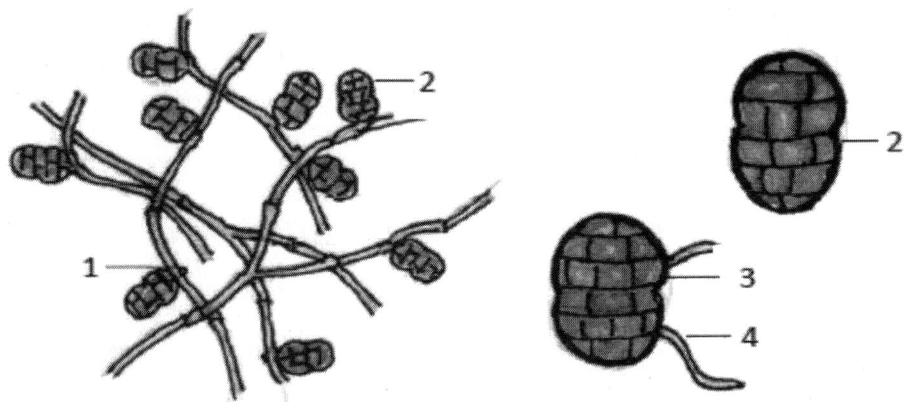

1. Conidiophore, 2. Conidium, 3. Germinating conidium, 4. Germ tube

Fig. 17. Type species - *Stemphylium vesicarium* (Leaf blight of garlic)

Life cycle of *Stemphylium*

The fungi belonging to this genus are mostly saprophytes and continue their lives in soil and infected plant debris and continue to produce conidia which cause the disease. The hyphae are septate, pale brown to brown in color. Conidiophores are simple or branched. They bear a number of vesicular swellings or nodes. Conidiogenous cells are terminally located. Conidia are solitary, light brown to black in color, rough or smooth-walled. They are muriform, oblong or sub-spherical and rounded at the tips. Conidia have transverse and vertical septations and a typical constriction is found at the middle. They have thickened scars at the base.

Species of *Stemphylium* causing diseases in spices - *Stemphylium vesicarium* on garlic and onion

18. Genus - *Puccinia* (Type species - *Puccinia psidii*)

The genus ***'Puccinia'*** is a large one comprising of about 4000 species distributed all over the world. All the species belonging to this genus are obligate plant pathogens and the disease caused by them is generally called as 'rust'. The bright yellow spores, which are produced in enormous numbers and spread over the surface of the affected parts impart an appearance of rust and hence the disease is called rust disease. Many species are most destructive to various economic plants.

The mycelium of the rust fungi is well developed and consists of septate, intercellular hyphae. The hyphae produce round or branched haustoria, which penetrate into the host cells and obtain nourishment Unlike most other Basidiomycota, the rust fungi do not produce basidiocarps.

Puccinia graminis f.sp. *tritici* causing black rust or stem rust of wheat; *P.graminis* f.sp. *hordei* causing rust of barley; *P. arachidis* causing rust of groundnut; *P. striiformis* causing yellow rust or stripe rust of wheat, barley and rye; *P.recondita* causing leaf rust or brown rust of wheat; *P.coronata* causing crown rust of oats; *P. sorghi* causing rust of corn; *P. purpurea* causing rust of sorghum are rust pathogens causing serious diseases in food crops.

Life cycle of *Puccinia*

The life cycle of the rust fungi is a complex and complicated one. There are three kinds of rusts viz., macrocyclic rusts, demicyclic rusts and microcyclic rusts. In a typical rust fungus there are five distinct reproductive stages in the life cycle. They are Stage 0 - Spermagonia bearing spermatia (otherwise called Pycnia bearing pycniospores) and receptive hyphae; Stage I - Aecia bearing aeciospores; Stage II - Uredinia bearing urediniospores; and Stage III - Telia bearing teliospores; and Stage IV - Basidia bearing basidiospores. While pycnia, aecia, uredinia and telia are formed on the host plant, basidia are formed outside the host. The telial stage is considered as the perfect stage of the uredinales because it is in the teliospores that karyogamy and meiosis take place. Rusts, which typically exhibit all the five reproductive stages are called 'macrocyclic rusts.

Demicyclic rusts lack the uredinial stage, while the microcyclic rusts lack the basidial, pycnial and aecial stages. Rust fungi requiring a collateral host besides the main host for completion of their life cycle are called 'heteroecious rusts', while those that complete their entire life cycle in a single plant species are called 'monoecious' or 'autoecious' rusts.

Puccinia psidii is a microcyclic rust in which the spermagonia and aecia are not formed. Uredinia are amphigenous, yellowish in color, more common on the abaxial surface, subepidermal becoming erumpent. Urediniospores, which are the repeating spores are globose or ellipsoidal to ovoid, yellowish in color and finely echinulate. Germ pores are indistinct. Telia are subepidermal, erumpent, pulvinate and yellowish-brown in color. Teliospores are cylindrical to ellipsoidal with rounded apex and narrow below, thick at the apex, yellowish-brown in color, two-celled, constricted at the septum and pediculate.

Basidia are cylindrical, hyaline and four-celled, Basidia are produced from each cell of the teliospore, apically in the upper cell and laterally in the lower cell. Basidiospores are globose to pyriform, hyaline and smooth. The disease is spread only through the urediniospores.

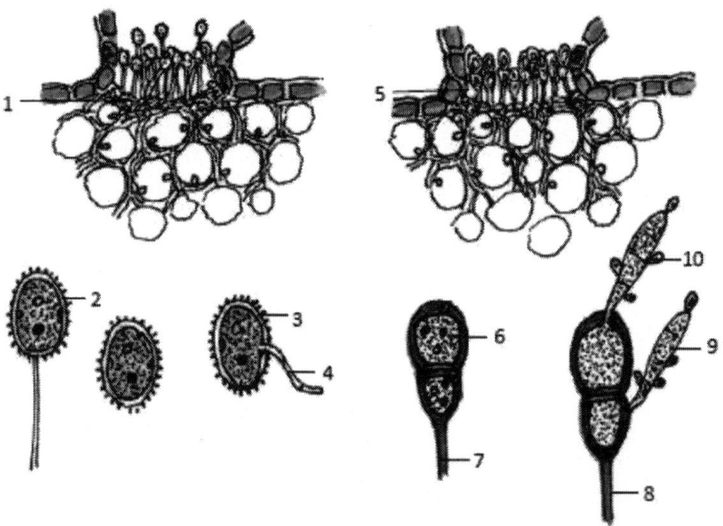

1. Uredinia, 2. Urediniospore, 3. Germinating urediniospore, 4. Germ tube, 5. Telia, 6. Teliospore, 7. Pedicel, 8. Germinating teliospore, 9. Basidium, 10. Basidiospore

Fig. 18. Type species - *Puccinia psidii* (Leaf rust of allspice)

Species of *Puccinia* causing diseases in spices - *Puccinia allii / P. porri* on garlic; *P. psidii* on allspice and clove; *P. pimpinellae* on aniseed; *P. nakarishikii* on lemon grass

19. Genus - *Erysiphe* (Type species - *Erysiphe polygoni*)

So called powdery mildew fungi belonging to Family Erysiphaceae are an important group of plant pathogenic fungi comprising of about 873 species in 17 genera, which infect a large number of species of angiosperms world wide and cause serious diseases of numerous economically important cultivated crops, as well as a wide variety of other plant species. All powdery mildew species are exclusively obligate plant parasites. The genus *Erysiphe* constitutes about 50% of the species causing powdery mildew.

Powdery mildew of coriander, legumes, beets and crucifers caused by *Erysiphe poligoni;* powdery mildew of begonia, chrysanthemum, cucurbits, dahlia, zinnia etc. caused by *E. cichoracearum;* powdery mildew of cereals and grasses caused by *E. graminis* f.sp.*tritici* are a few of the destructive diseases caused by species of the genus.

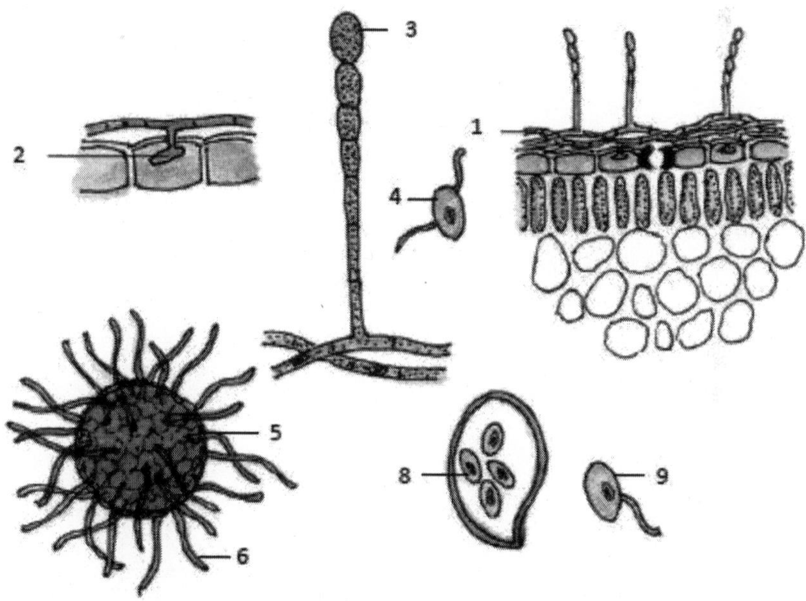

1. Mycelium 2. Haustorium, 3. Conidia, 4. Germinating conidium, 5. Cleisthecium, 6. Appendages, 7. Ascus, 8. Ascospores, 9. Germinating ascospores

Fig. 19. Type species - *Erysiphe polygoni* (Powdery mildew of coriander)

Life cycle of *Erysiphe*

The mycelium of these obligate parasites is ectophytic, septate and forms a fine net work of white, superficial hyphae on the host surface. The hyphae are attached to the host surface by means of small, round haustoria that mostly penetrate the epidermal cells. From the superficial mycelium erect, unbranched, septate conidiophores arise. Conidia are cut off at the ends of the conidiophores, one at a time in succession from below. Rarely conidia are formed in chains. Enormous number of conidia produced by the fungi and deposited on the host surface appear to the naked eye as a white, powdery coating and hence the disease is called as 'powdery mildew'. The conidia are hyaline, elliptical and single-celled. Cleistothecia, the sexual reproductive fruiting bodies produced later on are found scattered on the surface of the white mycelium, mostly in the diseased plant debris. They are black, globose and have mycelioid appendages. The appendages are variable in number (10 - 30) and length. They are invariably densely interwoven with the superficial mycelium. The cleistothecia contain 4 - 8 asci, which are ovate or subglobose and nearly sessile. Each ascus contains 3 - 5 ascospores, which are hyaline, elliptical and single-celled. The cleistothecia persist in the soil until the following season, when the wall disintegrates and the ascospores are liberated.

Species of *Erysiphe* causing diseases on spices - *Erysiphe polygoni* on coriander, fenugreek, cumin and tamarind; *E. heraclei* on aniseed; *E. cruciferarum* on mustard, black mustard and white mustard.

20. Genus - *Protomyces* (Type species - *Protomyces macrosporus*)

'Protomyces' is a genus of yeast-like fungi that is currently defined as plant pathogens of only plants belonging to Family Umbelliferae and Compositae. Just over 10 species of *Protomyces* have been officialy accepted. All known species are plant parasites and they form galls on stems, leaves and fruits.

Life cycle of *Protomyces*

The pathogen is endophytic and found only in the tumours. The hyphae are inter-cellular, closely septate, branch irregularly, quite broad and are diploid. Some of the cells in the hyphae swell and form elliptical or globose bodies, that develop into chlamydospores. The chlamydospores, which are the resting bodies are multinucleate and the nuclei are diploid. They are surrounded by a thick exospore. The resting spore germinates in water and the inner wall is pushed through the outer wall and forms an elongated, more or less cylindrical sac or vesicle. The protoplasm from the spore passes into the sac and the multinucleate protoplasm is displaced to the periphery of the sac. Each of the diploid nucleus develops into an ascus. Meiotic and mitotic divisions of the nuclei take place in the sac. The four haploid nuclei formed from each of the diploid nucleus become four ascospores inside a wall-less ascus. The large number of wall-less asci thus formed in the sac makes it a 'compound spore sac' or 'Synascus'. On maturity, the ascospores separate and are forcibly ejected from the spore sac in a single mass. After liberation, the uninucleate, haploid ascospores reproduce further by budding and they may conjugate in pairs and become zygotes. The zygote germinates and forms a diploid mycelium that penetrates into the host by producing a germ tube and causes fresh infection.

Species of *Protomyces* causing diseases on spices - *Protomyces macrosporus* on coriander

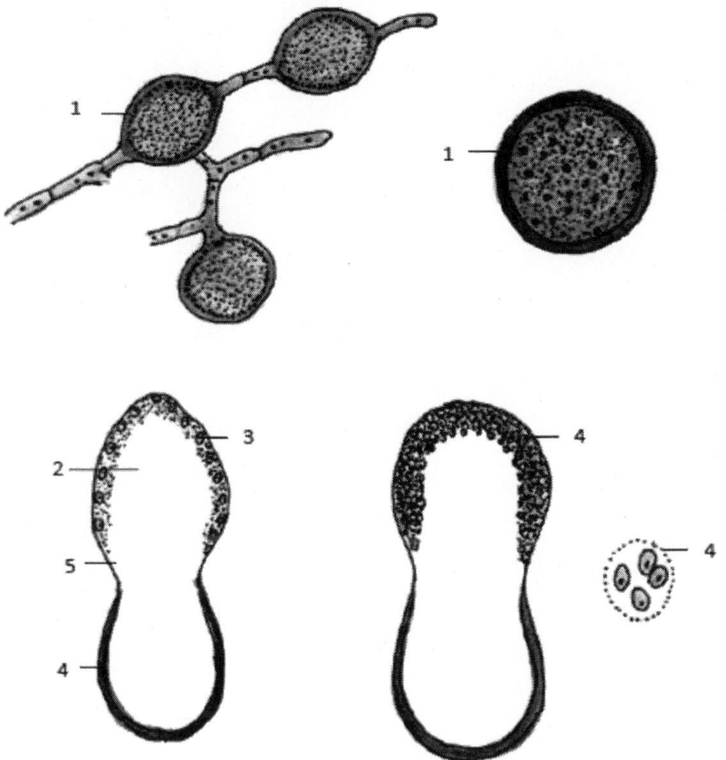

1. Chlamydospore, 2. Vesiclw, 3. Ascospores, 4. Outer wall, 5. Inner wall

Fig. 20. Type species - *Protomyces macrosporus* (Stem gall of coriander)

21. Genus - *Sclerotinia* (Type species - *Sclerotinia sclerotiorum*)

Genus '***Sclerotinia***' contains 14 species and are distributed widely all over the world. Stem rot disease caused by species of *Sclerotinia*, generally known as 'white mould' affects hundreds of plant species including many economically important crops.

Stem rot of many important crops, such as mustard, chillies, gingelly, coriander, lady's finger, egg plant, gram, sunflower, tobacco, tomato, potato, soybean, peas etc., cottony rot of carrot and watery soft rot of bean, cucumber etc. caused by

Sclerotinia sclerotiorum; peach fruit rot caused by *S. fructicola;* many destructive diseases of numerous succulent plants, particularly vegetables and flowers caused by *S. minor* are important diseases caused by species of the genus.

Life cycle of *Sclerotinia*

The mycelium of the fungi is hyaline, much branched consisting of large, closely septate hyphae, which are both inter- and intracellular. They invade all the tissues of the affected region. No true conidia are formed, however microconidia are formed exogenously and endogenously on short stalks. The microconidia are cut off in loose chains and they lie embedded in a slimy substance. But neither the microconidia nor the hyphae serve as a means of vegetative propagation. Once the vegetative growth has ceased, the hyphae collect in small dense masses, which gradually become the sclerotia. The sclerotia are whitish or pinkish at first and later turn black and become smooth. The smaller sclerotia are usually round, while the bigger ones are flattish and somewhat irregular in shape. The sclerotia are composed of thin-walled, rectangular cells in the center and thick-walled cells at the periphery. The walls of the peripheral cells are impregnated with a dark, gelatinous material, which gives the sclerotia a black appearance.

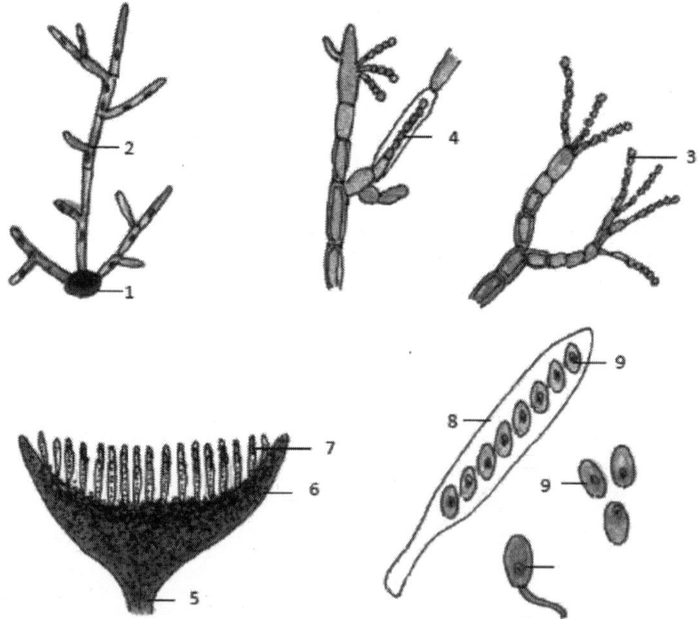

1. Sclerotium, 2. Mycelium 3. Exogenous microconidia, 4. Endogenous microconidia, 5. Stipe, 6. Apothecium, 7. Asci, 8. Ascus, 9. Ascospres, 10. Germinating ascospores

Fig.21-Type species - *Sclerotinia sclerotiorum* (Stem gall of coriander)

The sclerotia germinate during winter. Two to three stipes are produced and the apothecial fundamentals are borne at the tips of the stipes as minute, brownish, funnel-shaped structures just above the soil surface. Thc fundamentals expand into shallow, flat or concave apothecia, which become darker in color with age. The asci, which are arranged in a compact manner in the apothecia are cylindrical and have 8 ascospores in each ascus. The ascospores are hyaline and elliptical. Infection occurs only through the ascospores.

Species of *Sclerotinia* causing diseases on spices - *Sclerotinia sclerotiorum* on coriander, mustard, black mustard,and white mustard; *S. sclerotiorum*f.sp. *carotovora* on chicory

22. Genus - *Curvularia* (Type species - *Curvularia lunata*)

The genus ***'Curvularia'*** comprises of about 80 species of mould fungi, which are facultative pathogens on plants, animals and humans. They are found commonly in soil and debris. They are more in the tropical regions, while a few are found in the temperate regions also. Species of *Curvularia* cause diseases, such as grain mould, black kernel, seedling blight, leaf blights, leaf spots etc. *Curvularia lunata* is the most common and important species that causes grain mould and grain discoloration of rice and cereals, coriander etc., leaf spot of sorghum, blight disease of rice etc. The affected grains turn black and powdery and are unfit for consumption.

Leaf spot of coconut caused by *Curvularia maculans;* leaf spot of oil palm and grain discoloration of rice caused by *C. lunata;* leaf spot of tobacco and betel vine caused by *C. verrucolosa*; leaf blight of lemon grass caused by *C. andropogonis* are some of the other important diseases caused by species of *Curvularia.*

Life cycle of Curvularia

The type species *Curvularia lunata* appears as shiny, velvety-black, fluffy growth on the colony surface. The hyphae are septate and darkly pigmented. The hyphae produce brown, geniculate conidiophores. The conidiophores are erect, straight to flexuous, septate, often geniculate and sometimes nodulose. They produce conidia in sympodial succession. Conidia are elliptical, often curved or lunate, rounded at the ends, pale brown to dark brown and 3 - 10, mostly 3 - 5 septate. The basal and apical cells are pale brown in color, while the other cells are brown or dark brown, smooth-walled, curved at the third cell from the basal cell and larger than the other cells.

Species of *Curvularia* causing diseases on spices - *Curvularia lunata* on coriander

1. Conidiophore, 2. Conidia, 3. Conidium

Fig. 22. Type species - *Curvularia lunata* (Grain mould of coriander)

23. Genus - *Uromyces* (Type species - *Uromyces trigonellae*)

The genus *Puccinia* has the greatest number of species and ***'Uromyces'*** is the second largest genus of rust fungi and includes more than 600 species distributed all over the world. All the species of *Uromyces* are obligate plant pathogens. They are known to attack a wide variety of monocot and dicot plants and cause extensive damage.

Rust of bean caused by *Uromyces appendiculatus;* rust of broad bean and lentil caused by *U. fabae;* rust of carnation caused by *U. caryophyllinus;* rust of gram caused by *U. ciceris arietini;* rust of bean caused by *U. phaseoli typica;* rust of cowpea caused by *U. phaseoli vignae;* rust of pea caused by *U. pisi* are serious diseases caused by species of the genus *Uromyces.*

Life cycle of *Uromyces*

The pathogen is an autoecious, macrocyclic fungus and all spore stages occur on the primary host plant. However, the pycnial and aecial stages are rarely found. In a typical macrocyclic life cycle, all the five spore stages viz., Stage 0 - spermagonia (pycnia) bearing spermatia (pycniospores) and receptive hyphae; Stage I - aecia bearing aeciospores; Stage II - uredinia bearing urediniospores (uredospores); Stage III - telia bearing teliospores and Stage IV - Basidia bearing basidiospores (sporidia) occur in a sequential order.

The wind-borne basidiospore that happens to land on the host germinates by producing a germ tube and enters the host by penetrating the epidermis. From that hypha, a stromatic mass of hyphae consisting of uninucleate compartments develop below the epidermis of the upper leaf surface of the host. From this mycelial mat, a flask-shaped spermagonium (pycnium) develops with an outer pycnial wall and a central cavity. Large number of closely packed, elongated, uninucleate sporogenous cells (spermatiophores) are derived from the cells of the spermagonial wall and fill the spermagonial cavity. These cells give rise to a series of uninucleate spermatia (pycniospores) basipetally. A number of slender, stiff, tapering periphyses also develop from the upper edge of the spermagonial wall. The tips of the periphyses push the host epidermis from below, rupture it and protrude through the opening (ostiole) they have made. Spermatia, which are produced in enormous numbers are also exuded through the ostiole in a droplet of nectar, a thick, sticky, fragrant, sweet liquid. From the upper part of the spermagonial wall receptive hyphae are produced along with the periphyses and extend out through the ostiole. The spermatia and receptive hyphae produced in spermagonia from the sporidia of one mating type **(+)** are all of the same mating type **(+),** while the spermatia and receptive hyphae produced in spermagonia from the sporidia of the opposite mating type **(-)** are all of the opposite mating type **(-).** Spermatia and receptive hyphae of the mating type are not compatible and no copulation takes place between them. The periphyses may also function as receptive hyphae (stage 0)

Transfer of spermatia to compatible receptive hyphae are mostly effected by insects, which are attracted to the nectar or by rain splash or wind. When a spermatium comes in contact with a receptive hypha of the opposite mating type, the receptive hypha acts as a trichogyne and receives the nucleus from the spermatium. The nucleus of the spermatium moves into the receptive hypha through a pore dissolved at the point of contact and migrates down the receptive hypha, thus initiates the dikaryotic phase (stage I) of the rust life cycle.

As the spermagonia are being formed on the upper surface of the leaves, aecial primordia are formed from the uninucleate primary mycelium on the under surface of the leaves. The receptive hyphae are connected to the mycelium forming the aecial primordium. The nucleus from the spermatium migrates down the receptive hypha through the septal perforations and reaches the cells of the aecial primordium, making them binucleate. Conjugate division of these dikaryotic nuclei and formation of cross septa results in the formation of a large number of binucleate (dikaryotic) cells. A number of such dikaryotic cells eventually form a well-defined group of sporogenous cells or aeciosporophores, all in close lateral contact with at the base of the aecial primordium and develop into an aecium. From the sporogenous cells a chain of binucleate aeciospores and disjunctor cells are formed with the oldest cell at the top. In most cases, the peripheral cells of the aecial base form a wall or peridium that surrounds the spore chains in the form of a cup. When the aecium matures, the spore chains break the roof of the peridium and the spores are liberated. The torn peridium forms a lip around the aecial cup.

Aeciospores landing on the leaves of the host produce uredinia or uridia, which is the next stage (stage II) in the life cycle of the rust fungi. The spores produced are called urediniospores or urediospores or uredospores Besides aeciospores, the uridiniospores also produce uredinia. Urediniospores constitute the repeating stage of the rust fungi, since several generations of spores may be produced in one crop season. The uredinial cells are formed sub-epidermally from dikaryotic mycelium originating from an aeciospore or urediniospore. A palisade of hyphal tips appears in the uredinial initial and from this layer the urediniospores are produced. The spores are formed from buds originating from sporogenous cells. A bud enlarges and is then divided by a septum into two cells. The upper cell enlarges into the spore proper, while the lower cell develops into a stalk-like structure called the pedicel. As the spores form in large numbers, they exert pressure on the host epidermis from below and rupture it. Matured urediniospores are dikaryotic and have a thick wall covered with minute spines. They are mostly globose to ovoid in shape and pedicillate.

Telial stage which is the next stage (stage III) in the life cycle of rust fungi follows the uredial stage. Binucleate, thick-walled, unicellular spores called teliospores or teleutospores are formed in the telia. Telia may be formed afresh or the old uredinia may be converted into telia. Teliospores are formed from the tips of binucleate cells of the telium. The spore is at first dikaryotic, but later karyogamy takes place and the spore becomes diploid and uninucleate. The spores are stalked and are produced singly. The spore germinates and gives rise to basidiospores.

Under favorable conditions the teliospore germinates by producing a promycelium through the germ pore at the apex (stage IV). The diploid nucleus then migrates into the promycelium, undergoes meiosis and produces four haploid nuclei that distribute themselves at almost equal distances in the promycelium. Septa formed between the nuclei divide the promycelium into four uninucleate cells. Each cell then produces a sterigma at the tip of which a basidiospore is formed. The nuclei then migrate into the basidiospores. The basidiospores are forcibly ejected into the air. When a basidiospore lands on the host, it germinates by giving rise to a germ tube and enters the host by penetrating the host epidermis.

Species of *Uromyces* causing diseases on spices - *Uromyces trigonellae* on Fenugreek

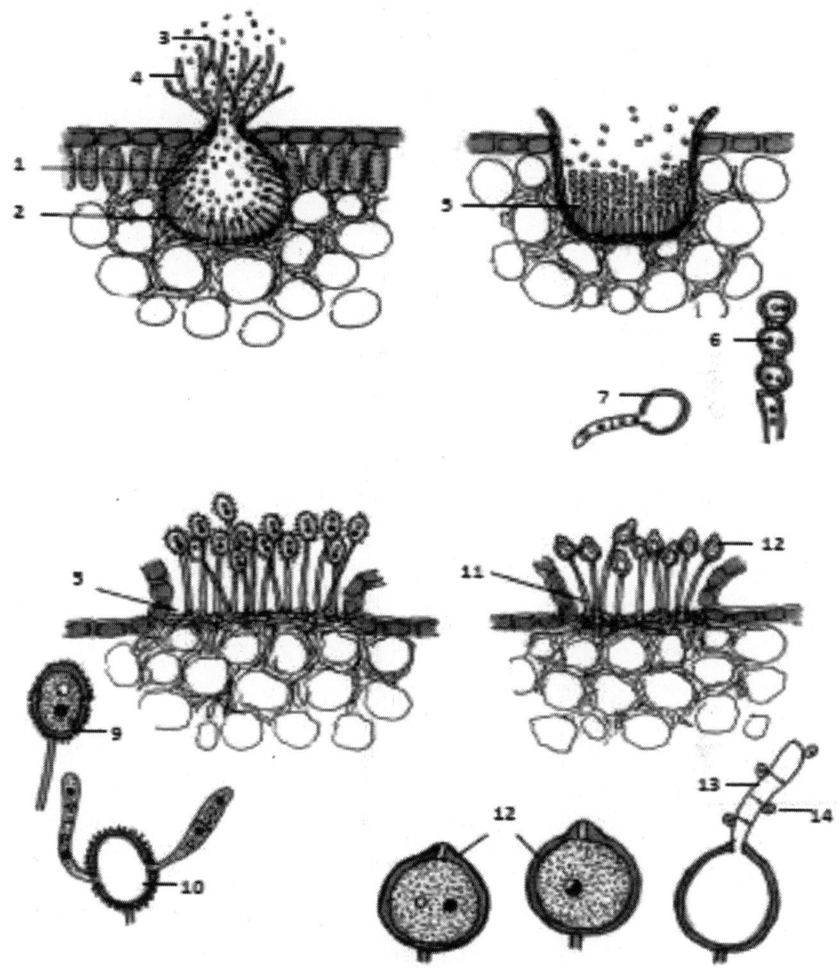

1. Pycnidium, 2. Spermationhores, 3. Pycniospores, 4. Receptive hyphae, 5. Aeoium, 6. Acciospores, 7. Germinating aeciospores, 8. Ureadinia, 9. Uredospore, 10. Germinating uredospore, 11. Telia, 12. Teliospore, 13. Basidium, 14. Basiospore

Fig. 23. Type species - *Uromyces trigonellae* (Rust of fenugreek)

24. *Albugo* (Type species - *Albugo candida*)

The genus ***'Albugo'*** is represented by 25 species and all the species are obligate parasites on land plants.

White rust of mustard, black mustard, white mustard, cabbage, cauliflower and other cruciferous crops caused by *Albugo candida;* white rust of egg plants caused by *A. bliti* are serious diseases caused by species of this genus.

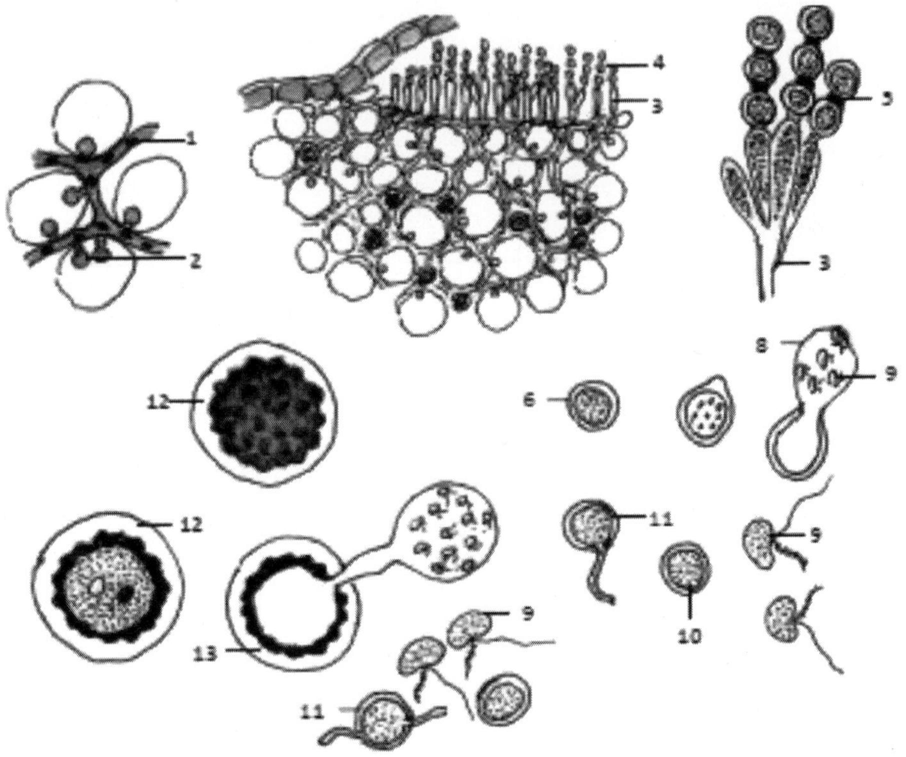

1. Intercellular mycelium, 2. Haustoria, 3. Sporangiophore, 4. Sporangia, 5. Isthmus, 6. Mature Sporangium, 7. Germinating sporangium, 8. Vesicle, 9. Zoospore, 10. Encysted zoospore, 11. Germinating encysted zoospore, 12. Oospore, 13. Germinating oospore

Fig. 24. Type species - *Albugo candida (Cystopus candidus)*- (White rust of mustard)

Life cycle of *Albugo*

The mycelium of *Albugo* is intercellular and feeds by means of globose or knob-like haustoria that penetrates into the host cells. Usually many haustoria are found inside a host cell. Hyphae from the endophytic mycelium congregate beneath the epidermis to form the sporangial bed. From the tips of large number of hyphal branches from the sporangial bed, short, club-shaped sporangiophores are produced. The sporangiophores are arranged in close proximity to one another in compact layers immediately below the host epidermis. The multinucleate sporangiophores produce a number of sporangia in succession from their tips in chains one below the other. The oldest sporangium is at the top of the chain and the youngest at the base. Pads of gelatinous material called 'isthmus' formed between each pair of sporangia function as disjunctors, which on disintegration free the sporangia. The growth of the mycelium and the continuous production of numerous sporangia exert pressure from below on the host epidermis, causing it to bulge and eventually rupture. Upon bursting of the epidermis, the sporangia are released and form a white crust on the surface of the host and so it is called 'white rust'. Sporangia are normally globose, but due to pressure during their formation some of them may be cuboid or polyhedral. They are somewhat thin-walled and multinucleate. The sporangia normally germinate by extruding 4-12 zoospores in a sessile vesicle, from which the zoospores emerge. Subsequently they encyst, germinate by putting forth a germ tube and infect the host. Under unfavorable conditions the sporangia germinate directly by putting forth a germ tube.

Sexual reproduction takes place as a result of fusion of the male nucleus from the antherudium and the female nucleus of the oogonium. Oospores, which are formed in the intercellular spaces of the hypertrophied parts of the host are thick-walled, dark yellowish-brown in color and have ornamental markings. The zygote (diploid) nucleus in the oospore undergoes meiosis followed by several mitotic divisions resulting in the formation of several haploid nuclei. The oospores germinate after a long period of rest by forming a vesicle at the end of a short tube and the zoospores are formed inside the vesicle. The zoospores encyst and germinate by germ tubes.

Species of *Albugo* causing diseases on spices - *Albugo candida* on mustard, black mustard and white mustard

25. Genus - *Plasmodiophora* (Type species - *Plasmodiophora brassica*)

The genus ***'Plasmodiophora'*** represented mainly by the species *Plasmodiophora brassicae,* the causal agent of club root disease is found worldwide and infects about 300 species in 64 genera of crucifers, both cultivated and wild. Economically important hosts include mustard, cabbage, cauliflower, broccoli, rape, kohlrabi, kale, Brussel sprouts, radish, turnip etc.

Life cycle of *Plasmodiophora*

Large number of resting spores found in the rotting tissues of galls of affected plants get access to the soil after disintegration of the host cells. The resting spores germinate and produce heterokont zoospores. The two, whiplash flagella of unequal length are inserted anteriorly. Larger resting spores, which contain a greater number of nuclei, produce a correspondingly larger number of zoospores. The zoospore finally comes to rest on a root hair, withdraws its flagella and surrounds itself with a delicate cell wall and forms a swarm spore. The protoplast escapes from the swarm spore and enters the cell of the host. In some cases the naked protoplast may penetrate the host cell without first forming a swarm spore. Within the host cell, the parasite develops into a multinucleate plasmodium. When the plasmodium reaches maturity, it becomes segmented into uninucleate portions, which become rounded in form. The nucleus divides two or three times and then each rounded portion, now multinucleate becomes enclosed by a delicate hyaline wall. This is regarded as a summer sporangium. The cytoplasm within the summer sporangium segments into uninucleate portions, which become pyriform, heterokont zoospores. The zoospores are released into the soil by disintegration of the host tissue. This asexual reproductive cycle is repeated.

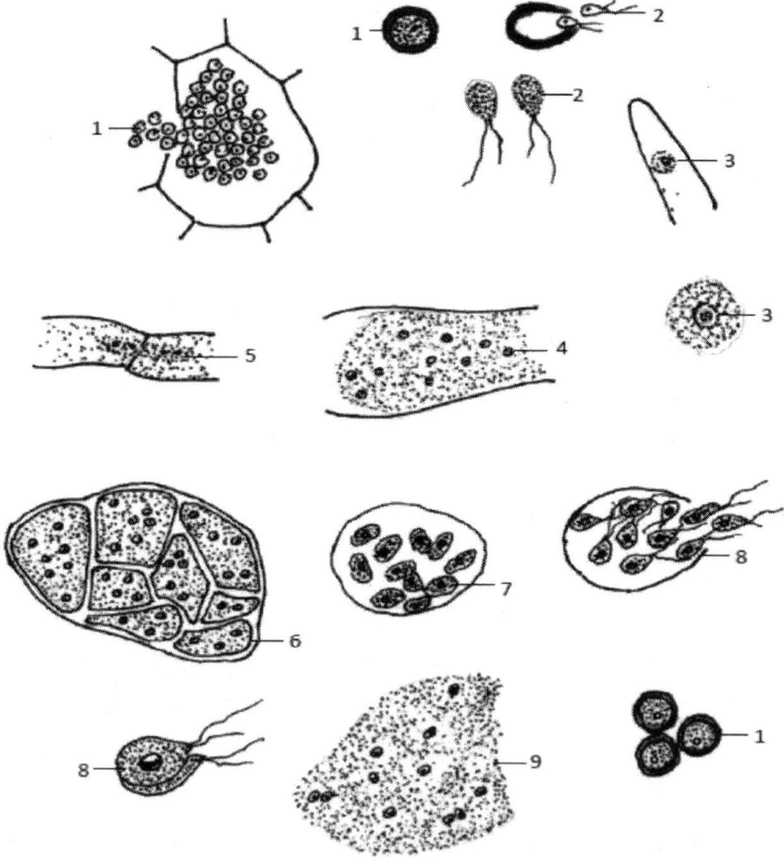

1. Resting spore, 2. Zoospore, 3. Swarm spore, 4. Plasmodium, 5. Summer porangium, 6. Segmentation of summer sporangium, 7. Formation of zoospores, 8. Copulation of zoospores, 9. Multinucleaate diploid plasmodium

Fig. 25. Type species - *Plasmodiophora brassicae* (Club root of mustard)

In the sexual reproductive phase, the zoospores function as planogametes and copulate in pairs. The young zygote then penetrates the epidermal cell of the host. After entry into the host, it develops into a multinucleate, diploid plasmodium. When reproductive maturity is attained, the nucleus undergoes meiosis by two nuclear divisions. The plasmodium then segments into uninucleate portions, which become rounded and are finally surrounded by a thin cell wall forming resting spores. Ultimately the resting spore germinates to form a zoospore.

Species of *Plasmodiophora* causing diseases on spices - *Plasmodiophora brassicae* on mustard, black mustard and white mustard

26. Genus - *Pestalotia* (Type species - *Pestalotia cinnamomi*)

The genus ***'Pestalotia'*** comprises of many important plant pathogens attacking mostly the foliage and fruits of various plants. and may cause extensive damage to the plants and yield loss. The pathogens are distributed worldwide.

Gray leaf spot or blight of coconut and oil palm caused by *Pestalotia palmarum*; fruit canker of guava caused by *P. psidii;* leaf spot of sapota and fruit rot of pomegranate caused by *P. versicolor;* gray blight or leaf spot of mango caused by *P. mangiferae;* gray blight of cashew caused by *P. microspora;* fruit rot of grapes caused by *P. menezesiana;* gray blight of tea caused by *P. theae;* leaf spot of several ornamental plants caused by *P.guipin;* gray leaf spot / blight of cinnamon caused by *P. cinnamomi* are some of the important diseases caused by different species belonging to this genus.

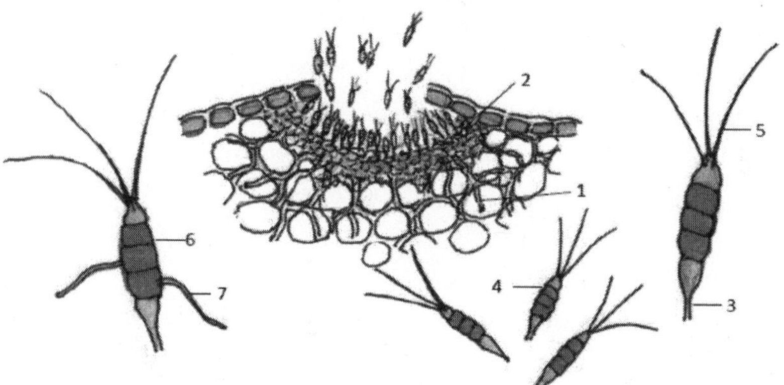

1. Mycelium 2. Acervulus, 3. Conidiophore, 4. Conidia 5. Appendages, 6. Germinating conidium, 7. Germ tube

Fig. 26. Type species - *Pestalotia cinnamomi* (Grey leaf spot of cinnamon)

Life cycle of *Pestalotia*

The mycelium of fungi belonging to this genera is septate, much branched and mostly intercellular and rarely intracellular. The intercellular mycelium produces haustoria, which enter the plant cells and absorb nutrients from the host. At the end of the vegetative phase, asexual reproductory phase commences. At this phase the hyphae aggregate below the host epidermis to form a stroma. The fruiting body or acervulus, a saucer-shaped structure is formed on the stromatic base. From the stroma crowded, minute, short, hyaline, unbranched, non-septate conidiophores arise. From the tip of the conidiophores, conidia are produced one by one in succession. The conidia are eye-shaped, with both the ends tapering towards the ends and are four-septate. The two terminal cells are hyaline, while the three central cells are dark-colored. The conidia are provided with three to five, mostly three slender, elongated appendages at the apex. The spores come out through the opening at the top of the acervuli. They germinate from one or more of the central cells by producing a germ tube.

Species of *Pestalotia* causing diseases in spices - *Pestalotia ccinnamomi* on cinnamon and cassia;

27. Genus - *Corticium* (Type species - *Corticium salmonicolor*)

The fungi belonging to the genus ***'Corticium'*** is generally called as 'Crust fungi'. The genus comprises of about 25 widely distributed species Some species of this genus are parasitic on wood or other economic crops, while some are saprophytes and live in the soil. Some species of this genus were formerly placed in the form genus *Rhizoctonia*.

Corticium salmonicolor causing 'Pink disease' of rubber, tea, coffee, cocoa, custard apple, cinnamon etc.; *C. koleroga* causing "Koleroga' or black rot on leaves and fruits of coffee; *C. fuciforme* causing 'Red thread disease' of turf grass; *C. solani* causing 'Black scurf' of potato and 'Bottom rot' of lettuce; *C. theae* causing 'Black rot' of tea are important pathogenic species in this genus

Life cycle of *Corticium*

The members of this genus produce very thin, webby or loose mycelium or a thick interwoven mycelium forming a crusty fructification overlying the substratum. These fruiting bodies or sporophores are formed on wood, herbaceous plant parts and sometimes even in the soil. The sporophores are often a little more than a thin layer of basidium bearing hyphae. The basidia develop on the crusty substratum. The basidia are single-celled, long, club-shaped and have four sterigmata at the apex. From each of the sterigma a basiospore is formed. The basidiospores are hyaline and oval-shaped. The basidiospore germinates by producing a germ tube and infects the host. The basidia are interspersed with sterile basidia and filamentous hyphal structures called 'cystidia'. These structures provide support to the basidia and keep them upright.

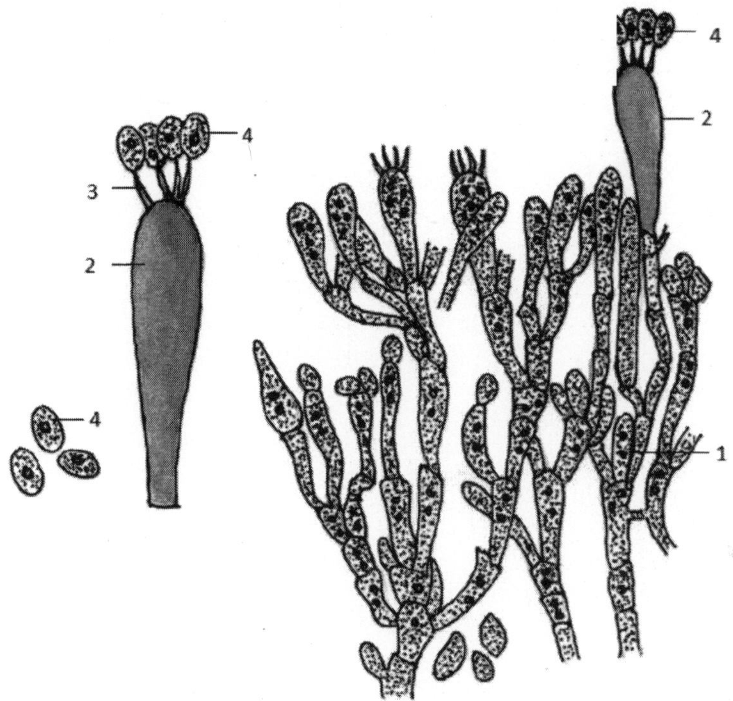

1. Sporophores, 2. Basidium, 3. Sterigma, 4. Basidiospores

Fig. 27. Type species - *Corticium salmonicolor* (Pink disease of cinnamon)

Species of *Corticium* causing diseases in spices - *Corticium (Erythricium) salmonicolor* on cinnamon and cassia;

28. Genus - *Ceratocystis* (Type species - *Ceratocystis fimbriata*)

The genus ***'Ceratocystis'*** consisting of more than 100 species includes many economically important tree pathogens world wide. Diseases caused by these fungi include vascular wilts, sap stains on logs and lumber, stem cankers and rots of roots, stems and fruits. Many species of *Ceratocystis* are primarily xylem pathogens. Infection occurs mainly through fresh wounds either natural by boring insects or from human activities. Infection may also occur through roots. The fungi cause dark, reddish-brown to deep brown or black staining in the xylem, which may extend from the roots up the trunks and into branches. On the surface of the trunk or branches cankers may develop, which may exude gummy substances. Most *Ceratocystis* species are well adapted to dispersal by insects.

Oak wilt caused by *Ceratocystis fagacearum;* 'Dutch elm disease' caused by *C. ulmi;;* black rot of sweet potato, blight of mango, mouldy rot of rubber, ficus wilt, wilt and die-back of eucalyptus etc. caused by *C. fimbriata;* 'Blue stain' in timber caused by *C. pilifera;* stem bleeding disease of coconut and 'Pine apple disease' or sett rot of sugarcane caused by *C. paradoxa* are important diseases caused by different species of this genus.

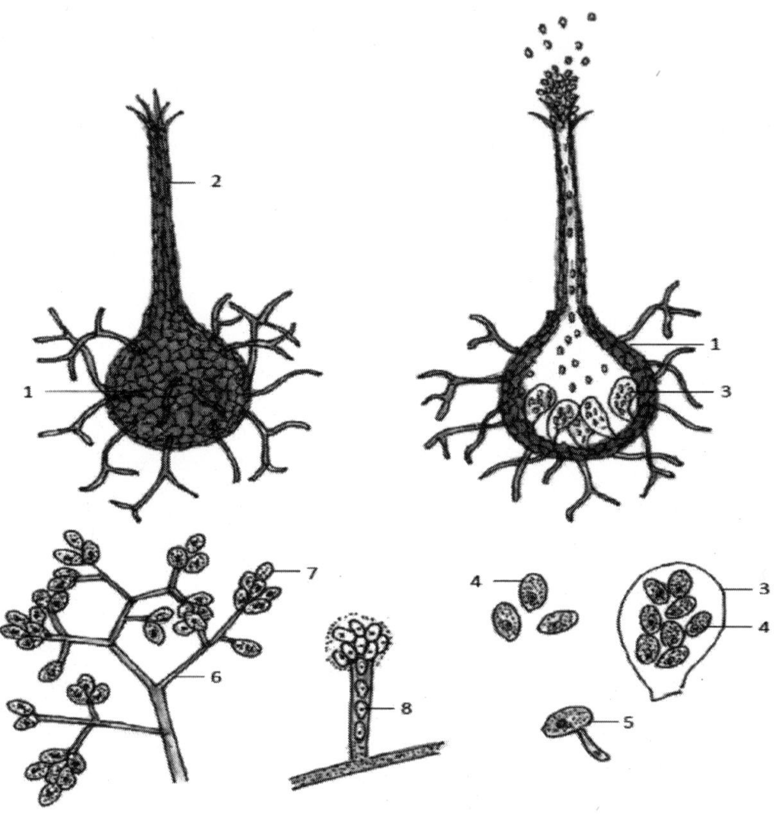

1. Perithecium, 2. Neck, 3. Ascus, 4. Ascospores, 5. Germinating ascospore, 6. Conidiophore, 7. Conidia, 8. Endogenous microconidia

Fig. 28. Type species - *Ceratocystis fimbriata* (Die-back / canker of allspice)

Life cycle of *Ceratocystis*

Perithecia, the fruiting bodies of the fungi are formed in dead wood or bark along the canker perimeter and are superficially or partially buried on the substrate. However perithecia are hard to detect. Perithecia have globose base and greatly elongated neck with shredded, feathery tip. The ascus wall gelatinizes early in the perithecial formation and the ascospores are extruded through the long neck of the perithecium and are embedded in mucus that forms a droplet at the ostiolar opening. Ascospore vary in shape from elongate to ovoid or crescent-shaped or hat-shaped. These fungi also produce conidia on short, simple conidiophores in chains one after another and are held together by a drop of mucus. The fungus also produces endogenous microconidia.

Species of Ceratocystis causing diseases in spices - *Ceratocystis fimbriata* on Allspice.

29. Genus - *Diplodia* (Type species - *Diplodia theobromae*)

'Diplodia' is a large genus with over 1000 recognized species. Species of *Diplodia* are known as pathogens on many woody hosts including fruit trees worldwide.

Crown wilt, dieback and cankers on pines caused by *Diplodia pinea;* black rot and cankers of apple caused by *D. mutila;* 'frog-eye' leaf spot, black rot and cankers of apple caused by *D. seriata;* cankers and dieback of oaks caused by *D. corticola;* sett rot of tapioca, gummosis in citrus, cucurbits and a number of other cultivated crops caused by *D. natalensis;* stem canker of red gram caused by *D. cajani;* boll rot of cotton caused by *D. gossypina;* dieback of roses caused by *D. rosarum;* fruit rot of nutmeg caused by *D. theobrome* are serious diseases caused by species of this genus.

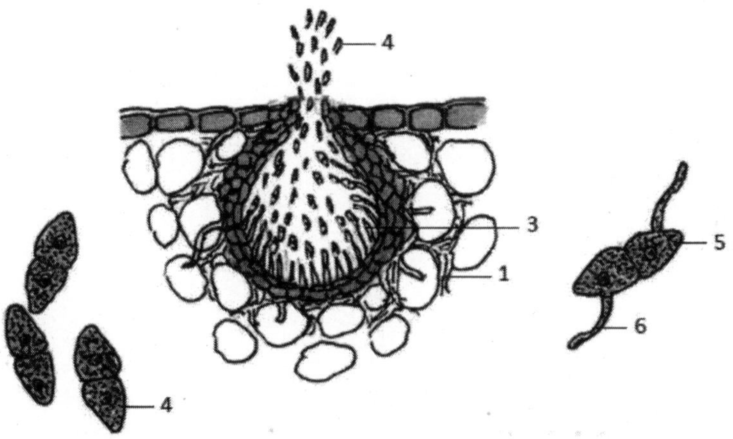

1. Mycelium, 2. Pycnidia, 3. Conidiophore, 4. Conidia, 5. Germinating conidium, 6. Germ tube

Fig. 29. Type species - *Diplodia theobromae* (Fruit rot of nutmeg)

Life cycle of *Diplodia*

The fungus enters the host either through wounds or by direct penetration of the host epidermis. The mycelium is inter- and intra-cellular. At the end of the vegetative phase the fungus produces asexual fruiting bodies (pycnidia). The pycnidia are erumpent, black, single and globose. Conidiophores are slender, simple and hyaline. They bear conidia at their tips one by one in succession. The conidia are dark, two-celled and ellipsoid or ovoid. Conidia, which are produced in large numbers come out of the pycnidia through the ostiole of the pycnidia during rainy periods and are dispersed by rain splash.

Species of *Diplodia* causing diseases on spices - *Diplodia theobromae* on nutmeg

30. Genus - *Botrytis* (Type species - *Botrytis cinerea*)

The genus ***'Botrytis'*** comprises of about 30 species and *Botrytis cinerea* is the most important one. While the genus contains many well known plant parasites, some of the species exist as saprophytes in various environments all over the world.

Grey mould of grapes, strawberry and some other fruits caused by *Botryris cinerea*; grey mould of onions and shallots in storage caused by *B. allii;* 'Fire' disease on living tulip leaves by *B. tulipae;* chocolate spot on broad bean by *B. fabae* are some of the serious diseases caused by species of this genus.

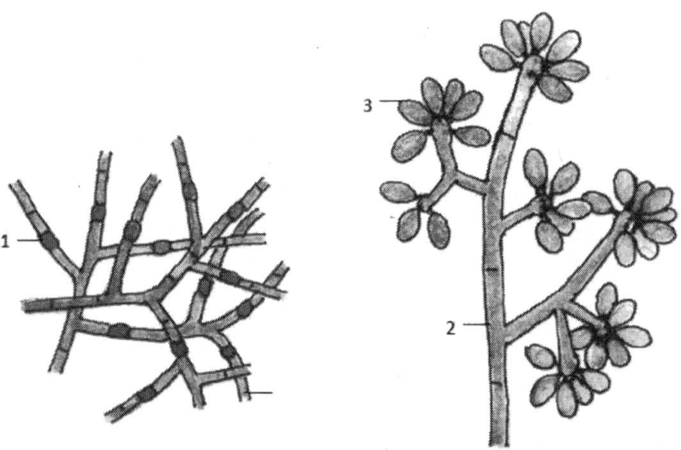

1. Mycelium, 2. Conidiophore, 3. Conidia

Fig. 30. Type species - *Botrytis cinerea* (Gray mould of basil)

Life cycle of *Botrytis*

The characteristic symptom of the fungus is the production of brown to grey, dense mycelial growth on the infected parts of the plants. From the weft of mycelium, grey, branching, tree-like conidiophores arise in large numbers. Hyaline conidia are borne at the tips of conidiophores, which appear as bunches of grapes. Asexually the fungus also produces highly resistant sclerotia, as well as chlamydospores. These resting spores can remain in the soil and in the debris for prolonged periods and initiate fresh infection. The enzymes produced by the fungus can dissolve the cell walls of the tissues and obtain nourishment from the cells, as a result the parts rot. Sexual reproduction is very rarely found and so the entire life cycle is repeated only by means of the asexual spores. The conidia are dispersed by wind or rain splash, while the resting spores are dispersed mostly by water coming from the rains or by irrigation water.

Species of *Botrytis* causing diseases on spices - *Botrytis cinerea* on basil; *B. allii* on onion

31. Genus - *Septoria* (Type species - *Septoria lactucae*)

'Septoria', a pycnidia producing fungi cause numerous leaf spot diseases on field crops, forages and vegetables including tomato and are responsible for causing heavy yield losses. The genus is wide spread and estimated to contain 1072 species. Pycnidia produce needle-like pycnidiospores. Most *Septoria* species are host specific and affect only a few plant genera. The fungus may remain in the infected plant debris for up to 2 years.

Leaf spot of tomato caused by *Septoria lycopersici*; leaf spot of chrysanthemum caused by *S. chrysanthemella;* blight of celery caused by *S. apii;* leaf blotch of wheat and other cereals caused by *S. tritici;* leaf spot of cowpea caused by *S. vignicola;* brown spot of soybean caused by *S. glycines;* leaf spot of sunflower caused by *S. helianthi;* blight of pea caused by *S. pisi;* concentric leaf spot of sweet potato caused by *S. bataticola;* leaf spot of rose caused by *S. rosae;* leaf spot of blackberry caused by *S. rubi;* leaf spot of pear caused by *S. pyricola* are important diseases caused by species of this genus.

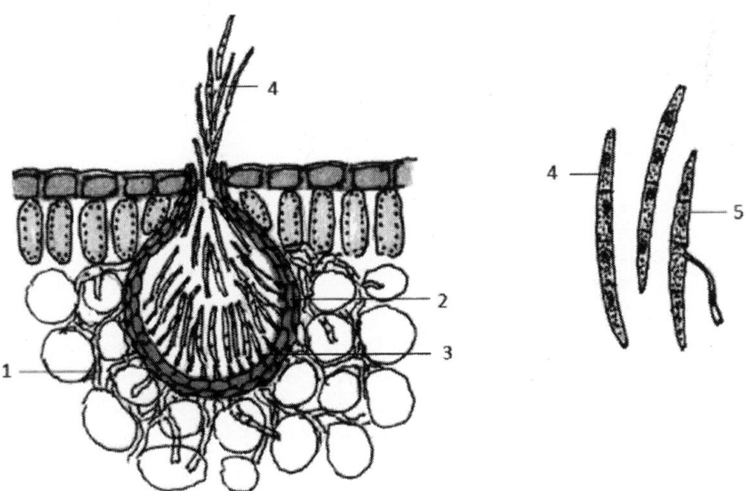

1. Mycelium, 2. Pycnidia, 3. Phialides, 4. Conidia, 5. Germinating conidium

Fig. 31. Type species - *Septoria lactucae* (*Septoria* blight of chicory)

Life cycle of *Septoria*

The mycelium of the fungus is well-developed, branched, inter- and intra-cellular, septate, hyaline when young and becoming slightly darker with age. Pycnidia, the asexual fruiting body of the fungus are minute, dark brown in color, globose, ostiolate and sunken in the host tissue. Conidia are borne at the tips of short, hyaline phialides in succession. They are hyaline, long, slender, filiform, 3 - 9 septate and slightly curved. When the pycnidia get wet, they swell and the conidia are extruded in long tendrils through the ostiole.

Species of *Septoria* causing diseases on chicory - *Septoria lactucae* on chicory

32. Genus - *Bremia* (Type species - *Bremia lactucae*)

Downy mildew caused by fungi belonging to Genus ***'Bremia'*** infects a few crops and *Bremia lactucae* causing downy mildew of chicory is the most important species.

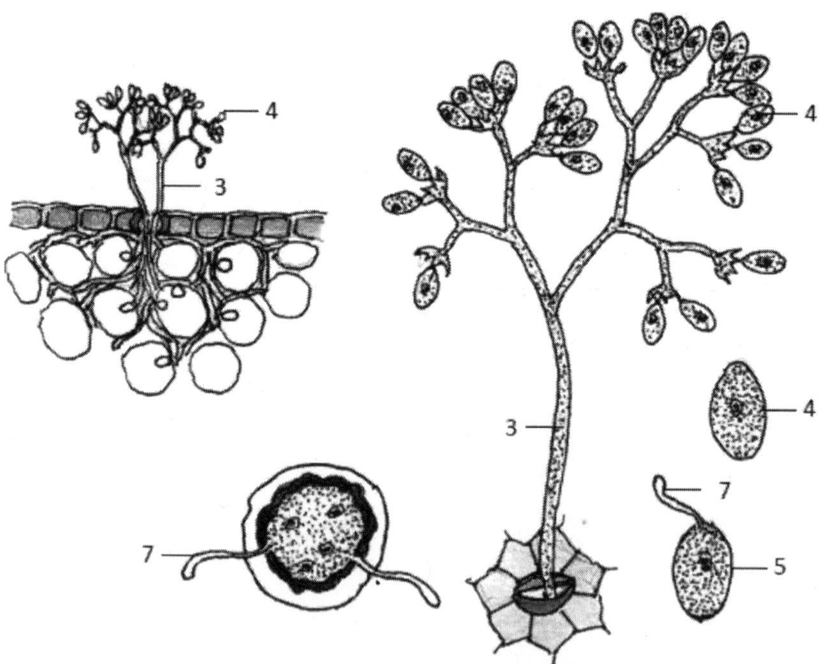

1. Mycelium, 2. Haustorium, 3. Sporangiophore, 4. Sporangium, 5. Germinating sporangium 6. Germinating oospore, 7. Germ tube

Fig. 32. Type species - *Bremia lactucae* (Downy *mildew* of chicory)

Life cycle of *Bremia*

Bremia lactucae is an obligate plant pathogen and it causes the serious downy mildew disease of chicory. The mycelium is coenocytic, hyaline and inter-cellular. Sac-shaped haustoria are produced from the hypae and haustoria may be present in almost all the cells. In the asexual reproductive phase, the fungus produces sporangiophores, which may emerge singly or in small groups through the stomata. The sporangiophores branch dichotomously. The tips of each branch expand to form a cup-shaped disc called 'apophyses' bearing mostly four, cylindrical sterigmata at the margin and from these sterigmata hyaline sporangia arise singly. Germination of sporangia is usually by means of germ tube from the apical papilla that forms an appressorium and penetrates the epidermal cell or it enters the host through a stoma. Sexual reproduction is oogamous the oospores are formed in the leaf tissue. The oospores are the resting spores and they remain viable for up to 12 months. Under favorable conditions when suitable hosts are present, the oospores germinate by means of a germ tube.

Species of *Bremia* causing diseases on chicory - *Bremia lactucae on chicory*

References

George N. Agrios (1997). Plant Pathology. *Academic Press,* San Diego, California.

Argunan G., Karthikeyan G., Dinakaran D., Raguchander T. (1999). Diseases of horticultural crops. *Department of Plant Pathology,* Tamil Nadu Agricultural University, Coimbatore.

Darwin Chjristdhas Henry L., Lewin Devasahayam H. (2011). Crop Diseases. New India Publishing Agency, Pitam Pura, New Delhi-110 088.

Remasri A.P., Srikumar V. Spice India Journal Published by Spices Board. *Print Express, Asoka Road, Sulur, Ernakulam.*

NET

Colour Plates

Chapter 2: Diseases of Spice Crops

Ginger Plants

Rhizome rot of ginger

Ginger rhizomes

Phyllosticta leaf spot of ginger

Yellow disease of ginger

Bacterial wilt of ginger

Sheath blight of ginger

Dry rot of ginger

Pepper Plants

Pepper Corns

Pepper Corns

Quick wilt / Foot rot of pepper

Pollu disease

Anthracnose of pepper

Slow wilt of pepper

Basal wilt of pepper

Stunt disease of pepper

Phyllody of pepper

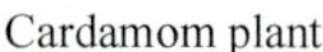

Cardamom plant

Cardamom pods

'Katte' disease of cardamom

Primary nursery leaf spot

Secondary nursery leaf spot

Capsule rot of cardamom

Leaf blight of cardamom

'Chenthal' disease of cardamom

Root tip rot of cardamom

Clump rot of cardamom

Nilgiri Necrosis of Cardamom

Cardamom vein clearing

Branches with buds

Dried clove

Canker of Clove

'Sumatra' disease of clove

Turmeric plants

Dried turmeric rhizomes

Rhizome rot of turmeric

Leaf blotch of turmeric

Colletotrichum leaf spot of turmeric

Garlic plants

Garlic bulbs

Damping off of garlic

Downy mildew of garlic

White rot of garlic

Bacterial soft rot of garlic

Purple blotch of garlic

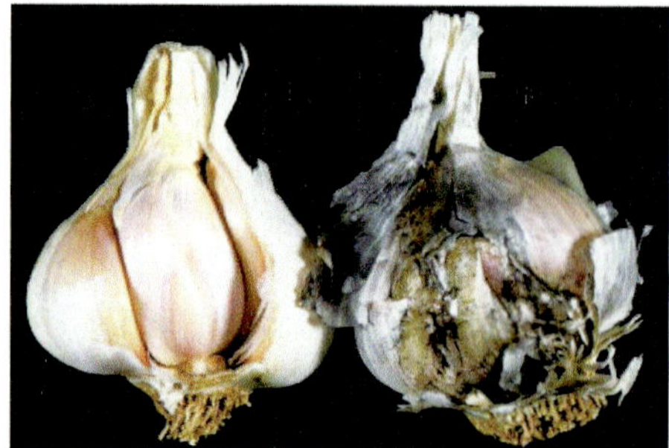

Black mould of garlic

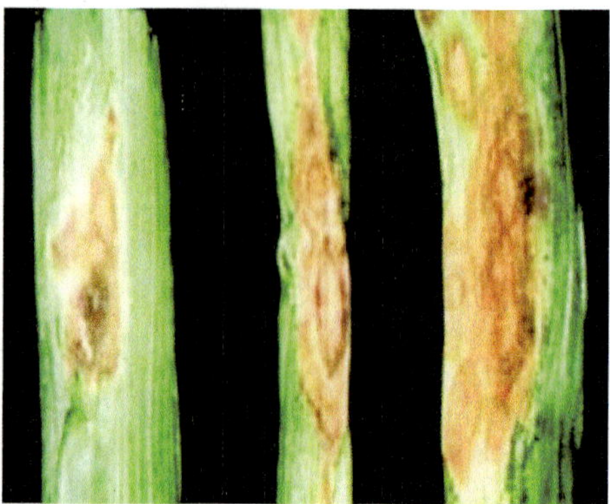

Stemphylium leaf blight of garlic

Onion yellow dwarf of garlic

Basal rot of garlic

Twister disease of garlic

Rust disease of garlic

Iris yellow spot of garlic

Powdery mildew of garlic

Coriander plants

Coriander seeds

Powdery mildew of coriander

Stem galls on stem

Stem galls on flowers and fruits

Wilt of coriander

Bacterial leaf spot of coriander

Fenugreek Plants

Fenugreek Seeds

Stem rot of coriander

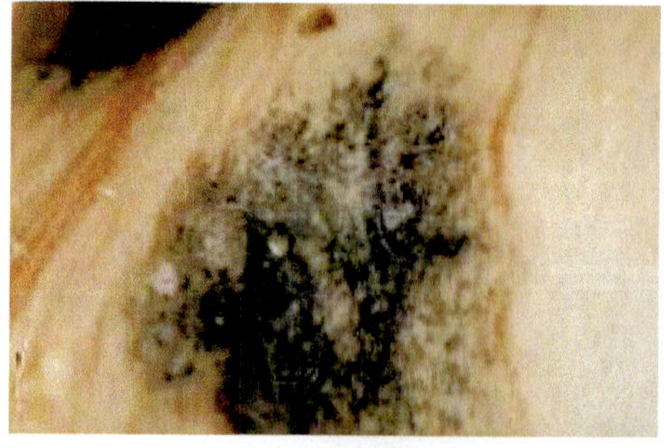

Cercospora leaf spot of fenugreek

Powdery mildew of fenugreek

Foot rot of fenugreek

Damping off of seedlings of fenugreek

Fusarium wilt of fenugreek

Cumin Plants

Cumin Seeds

Fusarium wilt of cumin

Powdery mildew of cumin

Alternaria blight of cumin

Black Cumin Plants

Black Cumin Seeds

Aniseed Plants

Aniseed Seeds

Damping off of cumin

Powdery mildew of Aniseed

Rust disease of aniseed

Mustard Plants

Mustard Seeds

Alternaria blight of mustard

Alternaria blight of mustard

White rust of mustard

White rust of mustard

Stag head-like appearance

Downy Mildew of Mustard

Downy Mildew of Mustard

Powdery mildew of mustard

Club root of mustard

Bacterial blight of mustard

Black mustard plants

Black mustard seeds

White mustard plants

White mustard seeds

Cinnamon tree

Cinnamon quills

Leaf spot of cinnamon

Stripe canker of cinnamon

Stripe canker of cinnamon

Grey Blight of cinnamon

Cassia tree

Cassia quills

Allspice immature berries

Mature berries

Myrtle rust of allspice

Nutmeg tree

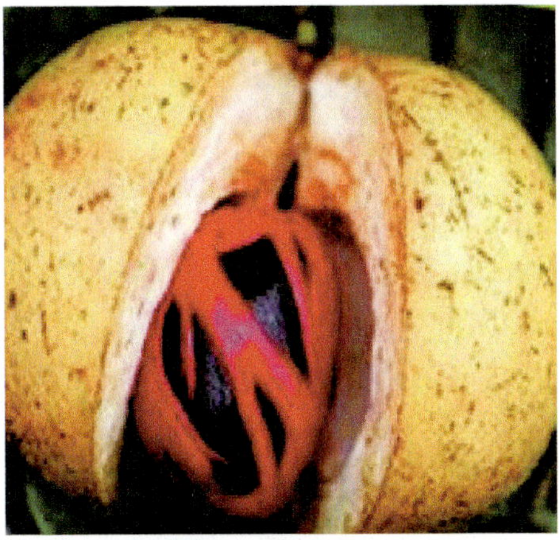

Nutmeg fruit

Leaf spot and shot hole of nutmeg

Fruit rot of nutmeg

Vanilla plants

Vanilla plant with pods

Dry vanilla pods

Vanilla seeds

Stem rot of vanilla

Shoot tip rot of vanilla

Stem blight of vanilla

Bean rot of vanilla

Gray mould of basil

Fusarium wilt of basil

Cercospora leaf spot of basil

Downy mildew of basil

Damping off of basil

Basil shoot blight

Fennel plant

Fennel flowers

Fennel seeds

Leaf blight of fennel

Leaf blight of fennel

Damping off of fennel

Powdery mildew of fennel

Saffron plants

Corms of saffron

Saffron flowers

Saffron

Corm rot of saffron

Chicory plants

Chicory roots

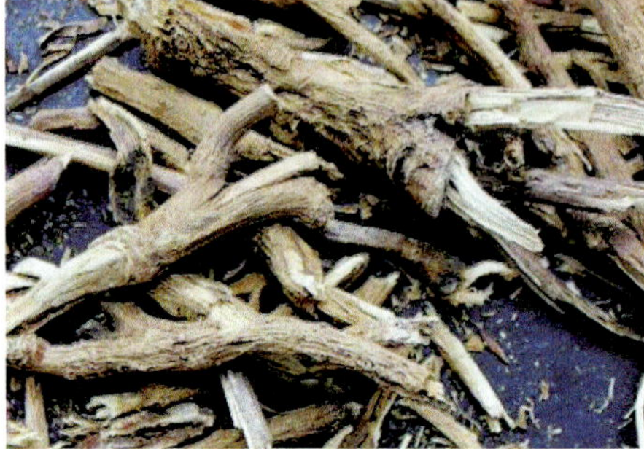
Dried chicory roots

Downy mildew of chicory

Septoria blight of chicory

Phyllosticta leaf spot of curry leaf

Lemon grass

Rust disease of lemon grass

Tamarind tree

Tamarind fruit

Powdery mildew of tamarind

Red chilli plants

Red chilli fruits

Damping off of seedlings of red chilli

Die back of chilli

Alternaria leaf spot of red chilli

Phytophthora blight

Fruit rot of chilli

Powdery mildew of red chilli

Frog-eye leaf spot of red chilli

Fusarium wilt of red chilli

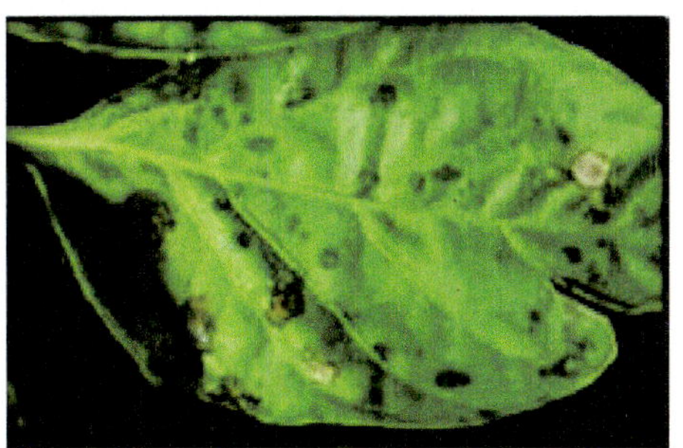

Bacterial leaf spot of red chilli

Mosaic disease of red chilli

Green chilli plant

Green chilli fruits

Asafoetida plants

Asafoetida resin